Fluid-Particle Technology: Analysis and Applications

Elmer L. Gaden, Jr., Series Editor

Alan W. Weimer, Volume Editor

John C. Chen, Liang-Shih Fan and Wen-Ching Yang, Volume Co-Editors

E. Aksoy
P. Basu
A.E. Bayly
X.X. Cheng
C.S. Daw
S. Dutta
J.S. Halow
J. Hinderer
C.-T. Huang
Y. Itani
R. Jackson
F. Keil
M. Kikuchi
H.O. Kono
L.-P. Leu
J. Li
L. Reh
C.D. Rielly
M. Sato
H. Schoenfelder
Y. Shibuya
J. Talukdar
H. Tanner
A. Tsutsumi
R. Turton
J. Wein
J. Werther
S. Wu
K. Yoshida

AIChE Staff
Maura N. Mullen, Managing Editor, Jocelyn Kimmel, Editorial Assistant
Cover Design: Joseph A. Roseti

Inquiries regarding the publication of Symposium Series Volumes should be directed to:
Dr. Elmer L. Gaden, Jr., Series Editor
University of Virginia, Department of Chemical Engineering
Thornton Hall, Charlottesville, VA 22903-2442. FAX (804) 924-6270

AIChE Symposium Series
1994

Number 301 Volume 90

Published by
American Institute of Chemical Engineers

345 East 47 Street New York, N.Y. 10017

AIChE shall not be responsible for statements or opinions advanced in papers or printed in its publications.

Library of Congress Cataloging-in-Publication Data

Fluid Particle Technology : analysis and applications / Alan W. Weimer volume editor, John C. Chen, Liang-Shih Fan, and Wen-Ching Yang, volume co-editors
Includes index.
ISBN 0-8169-0652-1
1. Fluidization -- Congresses. I. Weimer, Alan Wesley.
II. Series: AIChE symposium series ; no. 301
TP156. F65F55 1994
660'. 284292--dc20 94-28117
CIP

Foreword

This volume of the AIChE Symposium Series on Fluidization and Fluid-Particle Systems includes twelve peer reviewed papers selected from those presented in sessions at the AIChE Annual Meeting in St. Louis, MO, November 7th through the 12th, 1993. Included is the plenary paper "Progress Toward a Mechanics of Dense Suspensions of Solid Particles," as presented by Professor Roy Jackson.

The fluid-particle systems covered in this volume include dilute phase circulating and turbulent fluidized beds, dense phase vertical and centrifugal bubbling beds, vibrated fluidized beds, and coating processes. Fundamental considerations include the expansion behavior of turbulent beds, the measurement of rheological parameters and characterization of the emulsion phase in bubbling beds, bubble trajectories and void fractions in the flow around bluff bodies, chaotic phase-space trajectory characterization, and radial profiles of the slip velocity between gas and solids in circulating fluidized beds.

We acknowledge the outstanding work of the chairmen and co-chairmen of the sessions for their planning and paper selection. We also acknowledge the reviewers whose valuable comments provided needed guidance for the publication of this volume.

Alan W. Weimer, *Volume Editor*
Ceramics & Advanced Materials
Dow Chemical U.S. A.
Midland, MI

John C. Chen, *Volume Co-Editor*
Department of Chemical Engineering
Lehigh University
Bethlehem, PA

Liang-Shih Fan, *Volume Co-Editor*
Department of Chemical Engineering
Ohio State University
Columbus, OH

Wen-Ching Yang, *Volume Co-Editor*
Science & Technology Center
Westinghouse Electric Corporation
Pittsburgh, PA

Acknowledgement

The editor thanks the following reviewers whose valuable comments provided needed guidance for publication of this peer reviewed volume:

Dr. J.C. Chen, Lehigh University, Bethlehem, PA (USA)
Dr. Y.-M. Chen, Shell Development Company, Houston, TX (USA)
Dr. R. Clift, University of Surrey, Guildford, Surrey (UK)
Dr. R.A. Cocco, Dow Chemical USA, Midland, MI (USA)
Dr. R.S. Dixit, Dow Chemical USA, Midland, MI (USA)
Dr. A. Dutta, GTE Products Corporation, Danvers, MA (USA)
Dr. L.-S. Fan, Ohio State University, Columbus, OH (USA)
Dr. I.H. Farag, University of New Hampshire, Durham, NH (USA)
Dr. M. Flytzani-Stephanopoulos, MIT, Cambridge, MA (USA)
Dr. G.D. Groninger, Dow Chemical USA, Midland, MI (USA)
Dr. R. Jackson, Princeton University, Princeton, NJ (USA)
Dr. H.O. Kono, West Virginia University, Morgantown, WV (USA)
Dr. E.K. Levy, Lehigh University, Bethlehem, PA (USA)
Dr. W. Nowak, Nagoya University, Nagoya (JAPAN)
Dr. M. Rhodes, University of Bradford, Bradford (UK)
Dr. A.Z. Szeri, University of Pittsburgh, Pittsburgh, PA (USA)
Dr. H. Tanner, Swiss Federal Institute of Technology, Zurich (SWITZERLAND)
Dr. A.G.I. van der Ham, University of Twente, Enschede (THE NETHERLANDS)
Dr. A.W. Weimer, Dow Chemical USA, Midland, MI (USA)
Dr. W.-C. Yang, Westinghouse Electric Corporation, Pittsburgh, PA (USA)
Dr. J.G. Yates, University College, London (UK)

Contents

Plenary Paper

Progress Toward a Mechanics of Dense Suspensions of Solid Particles

Roy Jackson

Princeton University, Department of Chemical Engineering, Princeton, NJ. 08544

For practical purposes the behavior of fluidized beds and particle transport systems has been based on empirical correlations or simplified models of phenomena such as bubbling and clustering. Though differential equations of motion for these two-phase systems have been around for about thirty years, only a few approximate solutions have been found, which have contributed significantly to our physical understanding of phenomena such as stability and bubble motion, but have not been used to predict the performance of practical devices. However, recent advances in speed and memory of computers raise the possibility of exploiting, to a much greater extent, direct numerical solution of the underlying equations of motion. It promises to be possible to follow, in some detail, the complex fluctuations in velocity and bulk density that are so typical of most of these systems, and hence to analyze their performance at a level that is limited only by the accuracy of the proposed equations of motion.

The present state of our understanding of the mechanics of fluidized beds will be surveyed, as it bears on the form of the equations of motion, and some direct solutions of these equations will be introduced.

INTRODUCTION

Last year this talk was given by Roland Clift and was entitled "An Occamist Review of Fluidized Bed Modelling". Clift invoked the famous principle of "Occam's Razor" (which I will paraphrase as: "Choose the simplest explanation that fits the facts") to distinguish between learning models, which serve to provide physical insight into the fundamental causes of observed phenomena, and development-design models, which are used to translate limited experimental information into the prediction of a plant configuration to meet given specifications. The purpose of the design model is to reduce the need for costly experimentation at ever increasing pilot scales, so clearly it is of value only if the associated calculations can be carried out more easily and cheaply than the experimental tests they are meant to replace. Applying this criterion, Clift concluded that attempts at the direct solution of equations of motion for the fluid-particle mixture fall into the class of "learning models", because of the difficulty of evaluating certain parameters in the equations and the formidable amount of numerical computation required to simulate realistic pieces of process equipment containing fluidized beds. He went on to suggest that a more practical development-design model might be obtained by accepting the presence of bubbles, with known properties, within the bed, and simulating the motion of a cloud of bubbles in mutual interaction.

The Occamist principle is clearly correct, and the trick lies in using the razor to make the cut in the right place. Five years ago I would have agreed completely with Clift's choice: we had learned quite a lot about the mechanics of motion in fluidized suspensions by thinking about the form of their equations of motion and examining the nature of certain solutions, such as those representing the growth of small perturbations from the uniform state, or the motion of an isolated bubble, but solutions representing the behavior of complete circulation loops, or even an isolated fluidized bed, seemed to be impractical. However, in applying Occam's criterion we must always bear in mind that the difficulty and expense of large scale computation is not a constant, but a decreasing function of time; indeed, a function that has been decreasing very rapidly over the past few years. Consequently a few pioneers have been testing the waters by attempting direct solutions for complete pieces of process equipment. As an example of this Figure 1 (from Gidaspow and Therdthianwong, 1993) shows some results of a time-dependent simulation of the transient behavior of a complete, circulating fluidized bed loop containing crushed calcined limestone, 14 sec after the introduction of a gas stream containing sulfur dioxide at the bottom of the riser. A balance equation for the SO_2 was solved together with the mechanical equations for the gas-particle system. The left hand side of the figure shows the particle concentration as a function of position, while the right hand side shows the SO_2 concentration and reveals that there is a significant breakthrough of SO_2 into the exit gas. Similar simulations of alternative configurations were able to identify one for which SO_2 breakthrough was negligible.

The coarseness of the computational grid used in this work can be judged from the rather bizarre appearance of the sloping standpipe returning the solids to the foot of the riser, and the present reliability of the predictions may be affected by such computational limits, but we can be sure that these will be relieved dramatically over the next few years, and will not then be a serious obstacle to simulations. The potential benefits of direct solutions of this sort, compared with the usual semi-empirical "modelling" of the system, are great. In the example of Figure 1 the SO_2 breakthrough is associated with the asymmetric placing of the standpipe exit at the foot of the riser, and is eliminated when this is replaced by a symmetric arrangement. But such detailed differences of configuration are not captured by conventional modeling techniques, which are unable to predict their important consequences.

This is but one example of the potential advantages of treating chemical reactors by the direct solution of appropriate boundary value problems for the coupled differential equations representing fluid mechanics, diffusion and chemical reaction. Better known, perhaps, are the analyses of vapor deposition reactors pioneered by Jensen and his research group. What is new here is the extension of the same techniques to a two-phase system, where the equations of motion are less firmly established and considerably more complicated. We must, therefore, recognize that the old barriers to the construction of development-design models based on differential equations of motion are crumbling rapidly; soon the limiting factor is likely to be our understanding of the proper form for these equations, rather than the computational difficulty of solving them. I shall, therefore, sketch where we are at present in the quest for such an understanding, how we got here, and what remains to be done.

EARLY WORK ON STABILITY AND BUBBLE MOTION

From their earliest applications it was recognized that gas fluidized beds of solid particles are usually traversed by rising voids, largely free of particles, which burst through the free surface of the bed in a manner reminiscent of vapor bubbles breaching the surface of a boiling liquid. These voids were, therefore, called bubbles, and it was also recognized that they are absent from most liquid fluidized beds. Thus fluidized beds were classified as "aggregative" or "particulate", corresponding to the presence or absence of visible bubbles. Quite early Wilhelm and Kwauk (1948) proposed an empirical criterion to distinguish the two cases based on the Froude number u^2_{mf}/gd_p, where u_{mf} is the minimum fluidization velocity and d_p the particle diameter, with bubbling when this group is greater than unity. There were also speculations as to why the particles part to pass around the rising bubble, rather than simply falling through it, with the most popular explanation invoking an increase in particle concentration above the bubble, to a point where the assembly of particles has a yield strength and can form some sort of arch, rather like a structural arch of masonry.

Attempts to look in detail at the mechanics of the interacting gas and particle phases were initiated by Davidson's landmark analysis of the motion of a rising bubble (Davidson, 1961), in which he postulated a potential flow field for the particle velocity outside the bubble (assumed to be spherical), then deduced the flow field of the gas percolating through the moving porous medium represented by the particles. With this picture it immediately became obvious that bubbles large enough to rise faster than the interstitial velocity of the fluidizing gas in the uniform bed would be accompanied by a closed, toroidal vortex of gas, centered at the center of the bubble and larger than the bubble in diameter. The existence of this circulating gas "cloud" was quickly confirmed experimentally by Rowe and Partridge (1963), and there have been many subsequent observations of its shape, size and location relative to the bubble.

However, Davidson's analysis is not mechanically complete, since the particle motion is pre-specified; the origin of the forces which give rise to this motion is not part of the picture. A formally complete treatment of bubble motion, or indeed any other motion of a fluid-particle mixture, must rest on two momentum balances, one for each phase, and two corresponding continuity equations expressing the conservation of both fluid and solid material. Such sets of equations were tentatively formulated in the early 1960's by the present writer and others (van Deemter & van der Laan, 1961; Jackson, 1962, 1963a,b; Hinze, 1962; Pigford & Baron, 1965; Murray, 1965a,b). The form postulated at that time by the present writer is reproduced below:

$$\frac{\partial n}{\partial t} + \mathrm{div}\,(n\mathbf{v}) = 0 \qquad (1)$$

$$\nu\frac{\partial n}{\partial t} - \mathrm{div}[(1 - n\nu)\mathbf{u}] = 0 \qquad (2)$$

$$nm\left\{\frac{\partial \mathbf{v}}{\partial t} + (\mathbf{v}.\mathrm{grad})\mathbf{v}\right\} - n\nu\rho_f\left\{\frac{\partial \mathbf{u}}{\partial t} + (\mathbf{u}.\mathrm{grad})\mathbf{u}\right\}$$

$$+ \mathbf{i}n(m - \nu\rho_f)g - \beta(n)\,(\mathbf{u} - \mathbf{v}) = 0 \qquad (3)$$

$$\rho_f\left\{\frac{\partial \mathbf{u}}{\partial t} + (\mathbf{u}.\mathrm{grad})\mathbf{u}\right\} + \mathbf{i}\rho_f g + \mathrm{grad}\,p$$

$$+ \beta(n)(\mathbf{u} - \mathbf{v}) = 0 \qquad (4)$$

Here (1) and (2) are the continuity equations for the particle and fluid phases, respectively, while (3) and (4) are the corresponding momentum balances. n denotes the number of particles per unit volume, **u** and **v** are the fluid and particle velocities, respectively, m is the mass of each particle and ν its volume, while ρ_f is the fluid density and p the fluid pressure. β(n) is a drag coefficient relating the drag force per unit bed volume exerted between the fluid and the particles to their relative velocity, **i** is a unit vector pointing vertically upward, and g is the specific gravity force. The only forces taken into account are gravity, buoyancy, and a drag force proportional to the relative velocity of the two phases; there are no viscous effects of any kind associated with the shearing motions of the fluid or the particle assembly. All the dynamical variables are to be interpreted as local averages over regions containing many particles, but still small compared with the smallest significant length scale of the problem.

With these equations it proved easy to show that an infinite, uniformly fluidized bed is unstable, with the dominant instability taking the form of a rising pattern of bands of alternating larger and smaller bulk density (Jackson, 1962, 1963a). It was also possible to generate a partial solution for an isolated rising bubble, representing the motion of the fluid and particles approximately in the region above the equatorial plane of the bubble, and predicting the rise velocity (Jackson, 1963b). The bubble solution predicted the bulk density above the roof of the bubble to be somewhat *smaller* than that in the uniform bed, in contrast to earlier assumptions, and it showed that the parting of the particles to flow round the bubble can be attributed to forces exerted on them by the gas flowing outward through the roof of the bubble. The uniform bed stability analysis predicted the same type of instability for both gas and liquid fluidized beds, but with a marked difference in time scales, with much more rapid growth in gas fluidized beds, where the amplitude typically doubled while the wave advanced through only a fraction of its wavelength. It was, therefore, tempting to speculate that this contrast in degree of instability between the two cases is related to the presence of visible bubbles in gas fluidized beds and their absence from those fluidized by liquids. The credibility of the stability analysis, and hence the equations of motion, was enhanced by the observation, in water fluidized beds of glass beads, of just the sort of density waves predicted by the theory, first by Slis et al. (1959) and later, in more detail, by Anderson (1967). Perhaps the earliest photograph of these waves, taken by Anderson in a 1.5in diameter bed of 1mm glass beads fluidized by water, is reproduced in Figure 2.

There has been a regrettable tendency in recent years to regard the instability of a bed to small disturbances as implying the presence of bubbles, in the usual sense of that term, and hence to regard a criterion of stability against small disturbances as providing also a criterion for whether or not the bed will bubble. It cannot be overemphasized that the two are not synonymous; small amplitude instabilities need not develop into recognizable bubbles, nor does the fact that a bed is stable against small disturbances necessarily imply that bubbles, if introduced in some way, cannot survive. The mechanical link between initial stability and the ultimate fate of growing instabilities is a long and very interesting one, about which much more will be said later.

FURTHER DEVELOPMENT OF THE EQUATIONS OF MOTION AND THEIR CONSEQUENCES

One totally unsatisfactory feature of equations (1)-(4) is highlighted by the stability analysis, which predicts that the rate of growth of the unstable waves increases with decreasing wavelength, tending to infinity as the wavelength shrinks to zero. Apart from the fact that the equations eventually break down for very short wavelengths, the experimental observations of instability waves, referred to above, show a well-defined dominant wavelength which is many times the particle diameter. Clearly, what is missing from the equations is some dissipative mechanism, since any viscous effects will damp out the shortest wavelengths selectively because of the high rates of shear which accompany them.

The missing terms arise from interactions between the particles, either by direct contact or through induced motion of the intervening fluid, and from viscous effects generated by a non-uniform fluid velocity field. However, intuition provides an uncertain guide to the proper algebraic form of these terms, and there seems to be little chance of deriving them formally from the equations of motion of individual particles and the Navier Stokes equations, which describe in detail the motion of the fluid around and between them. Accordingly two approaches have been adopted, namely formal averaging of the point equations of motion for each phase over regions large enough to contain many particles, or direct postulation of the form of the equations of motion, guided by general principles of continuum mechanics and regarding the two phases as a pair of interpenetrating and interacting continua. The first approach, which was adopted by Anderson and Jackson (1967), appears at first sight to have a firmer physical basis, but this is largely illusory since the process of local averaging leaves terms whose explicit form must be found by some sort of intuitively based closure assumptions; in other words, it merely focuses our ignorance into certain terms in the equations. The equations of motion generated by spatial averaging, as found by Anderson and Jackson, take the following form:

$$\frac{\partial \varepsilon}{\partial t} + \nabla.(\varepsilon \mathbf{u}) = 0 \qquad (5)$$

$$\frac{\partial}{\partial t}(1 - \varepsilon) + \nabla.[(1 - \varepsilon)\mathbf{v}] = 0 \quad (6)$$

$$\rho_f \varepsilon \frac{D_f \mathbf{u}}{Dt} = \nabla.\mathbf{E}_f - \mathbf{f} + \varepsilon \rho_f \mathbf{g} \quad (7)$$

$$\rho_s (1 - \varepsilon) \frac{D_s \mathbf{v}}{Dt} = \nabla.\mathbf{E}_s + \mathbf{f} + (1 - \varepsilon)\rho_s \mathbf{g} \quad (8)$$

where ε denotes the void fraction, **u** and **v** are the local average velocities of the two phases, ρ_f and ρ_s are the densities of the fluid and solid materials, and **g** is the specific gravity force vector. $\mathbf{E}_f$ and $\mathbf{E}_s$ are effective stress tensors associated with the two phases, including correlation terms analogous to Reynolds stresses generated by the averaging process, and **f** is the fluid-particle interaction force per unit total volume. The notations D_f/Dt and D_s/Dt denote material time derivatives following the velocities **u** and **v**, respectively.

Equations (5) and (6) are explicit as they stand, but (7) and (8) are not closed unless expressions are given for the two stress tensors and the interaction force, and it is here that intuitive ideas must be brought to bear. Anderson and Jackson (1967) discussed this matter at length and adopted the following forms:

$$\mathbf{E}_f = [-p + \lambda(\varepsilon)\nabla.\mathbf{u}]\mathbf{I} + \mu(\varepsilon)[\nabla\mathbf{u} + \nabla\mathbf{u}^T - 2/3\,\nabla.\mathbf{u}\,\mathbf{I}] \quad (9)$$

$$\mathbf{E}_s = [-p_s(\varepsilon) + \lambda_s(\varepsilon)\nabla.\mathbf{v}]\mathbf{I} + \mu_s(\varepsilon)[\nabla\mathbf{v} + \nabla\mathbf{v}^T - 2/3\nabla.\mathbf{v}\,\mathbf{I}] \quad (10)$$

$$\mathbf{f} = (1 - \varepsilon)\nabla.\mathbf{E}_f + \varepsilon\beta(\varepsilon)(\mathbf{u} - \mathbf{v}) + (1 - \varepsilon)C(\varepsilon)\rho_f \frac{d}{dt}(\mathbf{u} - \mathbf{v}) \quad (11)$$

(9) and (10) assume that the stresses associated with the fluid and particle phases are related to the respective velocity fields like those of Newtonian fluids, with effective pressures, bulk viscosities and shear viscosities, as indicated. Note that the effective viscosities and the pressure associated with the particle phase are assumed to be specified constitutively, as functions of the void fraction, while the pressure associated with the fluid is left to be determined dynamically, as would be expected since the compressibility of the fluid is neglected. There is no real justification for these forms of the stress tensors, except that they are the simplest which introduce the desired dissipative effects into the equations. The first term in the expression (11) for the interaction force represents the buoyancy reaction of stress gradients in the fluid phase on the immersed particles, the second is the drag force generated by the relative motion, and the third is a guess at virtual mass effects. The drag force is indicated to be proportional to the first power of the relative velocity, but of course this should be replaced by a more appropriate function of **u** - **v** at high values of a Reynolds number based on the slip velocity and the particle diameter. After inserting (9), (10) and (11) into (7) and (8) various forms for the momentum balances can be obtained by linear combination, for example

$$\rho_f \frac{D_f \mathbf{u}}{Dt} = \nabla.\mathbf{E}_f - \beta(\varepsilon)(\mathbf{u} - \mathbf{v}) + \rho_f \mathbf{g} \quad (12)$$

$$\rho_s (1 - \varepsilon) \frac{D_s \mathbf{v}}{Dt} - \rho_f (1 - \varepsilon) \frac{D_f \mathbf{u}}{Dt} =$$

$$\nabla.\mathbf{E}_s + \beta(\varepsilon)(\mathbf{u} - \mathbf{v}) + (1 - \varepsilon)(\rho_s - \rho_f)\mathbf{g} \quad (13)$$

There has been a good deal of discussion of the proper form for the momentum balances. Unfortunately some of the sharpest controversies have been vacuous, arising from insufficiently precise definition of terms, or failure to recognize that the same physical phenomena can be introduced into the equations in different terms, leading to the same final result when all the terms are assembled. This is a pity, as there are certainly serious questions to be addressed.

Anderson and Jackson (1968) reworked the theory of stability using the above equations and found, once again that both gas and liquid fluidized beds are unstable. However, the presence of viscous terms now ensured that the maximum rate of growth was bounded, and occurred at a non-zero value of the wavelength. Some of their results for nominal 1mm glass beads fluidized by water are reproduced in Figure 3, which shows the value of ξ in the growth factor $\exp(\xi t)$ as a function of the wave number, for various values of parameters associated with the particle phase stress tensor. These should be compared with similar predictions for an air fluidized bed of the same particles, shown in Figure 4. It is seen that the ordinate scales of the diagrams differ by an order of magnitude, which is a good indication of the relative growth rates for liquid and gas fluidization. Of course, the results depend on the values taken for the various parameters appearing in equations (9) - (11). Of these, we can make a good estimate of the drag force from correlations such as that of Richardson and Zaki (1954), but it is difficult to devise independent experiments to provide information about the particle phase pressure p_s, the virtual mass factor C, and the effective viscosities λ, μ, λ_s and μ_s, though various experiments to measure "the viscosity of a fluidized bed" give some indication that one or other of the viscosities must be as large as several poise in typical dense beds. This is a serious criticism of equations of motion of the above type, and it is important to look more closely at the rheological properties to see if they can be formulated in terms of a small number of parameters that can be evaluated by independent experiments. All that could be said in the late 1960's was that values could be chosen which predicted speeds, dominant wavelengths and growth rates for the disturbances in fair agreement with a rather limited set of observations on water fluidized beds of glass beads (Anderson and Jackson, 1969), and which were not inconsistent with other measurements on fluidized beds, such as those of "effective viscosity". Later, and more extensive measurements by El Kaissy and Homsy (1976) led to similar conclusions.

THE INTERVAL OF UNIFORM FLUIDIZATION IN GAS FLUIDIZED BEDS

Having demonstrated the instability of uniform beds and provided approximate solutions for the motion of a fully developed bubble (Jackson, 1963b; Murray, 1965b), the obvious next questions to pose were whether the equations of motion could trace the growth of a small disturbance in a gas fluidized bed into a fully developed bubble, and whether they could explain why similar small disturbances in most liquid fluidized beds do not grow into bubbles. However, a different problem intervened to claim the attention of workers in the field, generate a great deal of controversy and, eventually, illuminate significantly our understanding of the mechanical behavior of fluidized beds. Specifically it was noted, contrary to the predictions from stability theory described above, that certain beds could be expanded well beyond the point of minimum fluidization, as usually defined, without any evidence of visible bubbling, or any other observable disturbances. This interval of apparently ideal, stable fluidization was found primarily in gas fluidized beds of rather small particles, of the order of 100μm diameter or less. Geldart (1973), in his now very well known classification scheme, referred to particles of this type as aeratable, or Group A. Shortly after Geldart's publication the discrepancy between the stability analyses and this observation was resolved, in principle, in a paper by Garg and Pritchett (1975) who published a stability analysis essentially identical to that of Anderson and Jackson (1968), but explored values of the particle phase pressure p_s much larger than had been thought reasonable in the earlier work. This revealed that a large particle phase pressure or, more precisely, a sufficiently large value of its derivative $|dp_s/d\varepsilon|$, would suffice to stabilize the uniform bed. Since the magnitude of this quantity would be expected to decrease as the void fraction increases, this would account for the existence of an interval of stable behavior, bounded above by instability. The question then remained, what physical mechanism could be responsible for an effective particle phase pressure of the requisite size?

A possible answer was outlined as early as 1973 by Rietema, who argued that cohesive forces at points of solid-solid contact would enable sufficiently small particles to form expanded structures while still remaining in contact with each other, and these structures would then exhibit elastic behavior. (This is analogous to the mechanism of flocculation in colloids.) A stability analysis on this basis (Mutsers and Rietema, 1977a) showed that this mechanism introduces a term formally identical with the stabilizing term identified by Garg and Pritchett, and arguments were advanced to justify a magnitude large enough to account for the width of the observed interval of stability. Rietema's mechanism would imply that the particle assembly should exhibit a yield stress throughout the interval of stability and should not, therefore, behave in a fluid-like manner. In particular, if the whole bed is tilted its exposed surface should not remain horizontal, but should tilt up to a certain limiting angle determined by the yield stress. Rietema demonstrated that this behavior is, indeed, observed experimentally. This view of the expanded bed also has other implications since, with its emphasis on solid-solid contact forces, it suggests that the rheological behavior of the particle phase should be that of a granular material, rather than a Newtonian fluid, as was pointed out by Clift in this lecture last year.

Despite their success in accounting for the interval of stable expansion Rietema's ideas encountered vigorous opposition. It was argued, for example, that a stabilizing mechanism based on contact forces would be very sensitive to factors which affect these forces, such as humidity, while the only particle properties in the apparently successful Geldart classification are the size of the particles and the densities of the solid and the gas. An alternative stabilizing mechanism of hydrodynamic origin had earlier been suggested by Verloop and Heertjes (1970), and this type of mechanism was revisited by Foscolo and Gibilaro (1984), who emphatically rejected the role of contact forces and claimed to provide a fully predictive criterion, able to distinguish between bubbling and non-bubbling beds, in the simple form

$$\frac{(gd_p)^{1/2}}{v_t}\left(\frac{\rho_s - \rho_g}{\rho_s}\right)^{1/2} < 0.56n(1-\varepsilon)^{1/2}\varepsilon^{n-1} \qquad (14)$$

where bubbling occurs if the inequality is satisfied. (This provides an example of a matter referred to above. (14) is actually a criterion for stability of a uniform bed against small perturbations, <u>not</u> a criterion for bubbling!) In equation (14) v_t is the terminal velocity of fall of an isolated particle in the gas, and n is the Richardson-Zaki exponent for the particles in question, which depends only on their Reynolds number when moving with velocity v_t through the gas. The physical origin of the stabilizing term proposed by Foscolo and Gibilaro has seemed obscure to many workers, but more recently Batchelor (1988) has offered an alternative mechanism for hydrodynamic stabilization based on the existence of random fluctuations in the particle velocities and linked to diffusion of particles. His stability criterion is strikingly similar in form to (14), namely

$$\frac{(gd_p)^{1/2}}{v_t}\left(\frac{\alpha}{2(1+\theta)}\right)^{1/2} < n(1-\varepsilon)\varepsilon^{n-1} \qquad (15)$$

where α and θ are quantities related to virtual mass effects and the particle phase pressure which can, in principle, be determined by independent measurements.

Further evidence in favor of stabilization by contact forces has been advanced by Rietema and co-workers (Mutsers and Rietema, 1977b; Rietema and Piepers, 1990) and has been summarized in a recent monograph (Rietema, 1991). Most recently Tsinontides and Jackson (1993) have reported careful measurements of bed expansion and pressure drop during fluidization-defluidization cycles of FCC catalyst with air, supplemented by high resolution measurements of bulk density with a gamma ray densitometer. Taken with the evidence of Rietema's group these leave no doubt that contact forces are the operative mechanism in the FCC system, and that the expanded bed is stable before the bubbling point because it has a non-vanishing yield stress. Striking evidence of this is provided by Figure 5 (from Tsinontides and Jackson, 1993) which shows the variation of solid fraction with height along the axes of beds at various degrees of expansion within the stable region. Each point represents an average density over a period of 15-20 min., and variations larger than about twice the size of the icons used to indicate the experimental points are significantly bigger than the limits of experimental precision. There are seen to be quite large variations in solids fraction, particularly in the middle of the range of stable expansion and these are robust, in the sense that they will survive unchanged for days provided the bed is not subjected to severe mechanical disturbance. No hydrodynamically stabilized dispersion could sustain such structures.

While the above results are specific to the particular systems studied, and do not preclude the possibility of hydrodynamic stabilization for other cases, they do emphasize the central role of particle-particle contact forces in determining the mechanical behavior of gas fluidized beds, vindicating the views of the late Professor Rietema. We will return to this later.

THE FATE OF DISTURBANCES BEYOND THE RANGE OF LINEAR STABILITY THEORY

After the digression of the last section, productive though it proved to be, we return to the central question of the ultimate fate of growing disturbances in fluidized beds. Recall that linear stability theory accounts only for a difference in time scales between the initial growth in gas and liquid fluidized beds. Since the early publications there have been a number of attempts to extend the theory to follow the growth further. Most of these have concentrated on non-linear effects in the growth of one-dimensional disturbances, represented by waves whose wave vectors are directed vertically upward, since these are the most unstable type of disturbance in the linear approximation, but there has also been some work on the development of two-dimensional structures.

The earliest attempt to take account of nonlinear effects was an analysis by Fanucci et al. (1979) using the method of characteristics applied to one-dimensional equations of motion without any viscous terms. Qualitative arguments put forward much earlier by Pigford and Baron (1965) suggested that waves which are sinusoidal in form at low amplitude should distort as they grow, in such a way that the transition from a dense region below to a less dense region above should steepen, while the reverse should occur for the transitions from less dense regions below to denser regions above. This conclusion was confirmed by examining the slopes of the characteristics, and numerical integration of the characteristic form of the equations showed how the distortion develops, indicating that the sharp transitions should eventually grow into shocks. Shortly after this work Liu (1982, 1983) re-examined the linear stability theory from the point of view of wave hierarchies and derived a linearized Burgers-Korteweg-deVries equation for the void fraction. He also tackled the nonlinearities in the one-dimensional equations by Fourier expansion of the solution, which leads to an open hierarchy of equations for the Fourier coefficients. By neglecting all but low order coefficients the hierarchy can be closed and hence an approximate solution, presumably more accurate than that given by the linear theory, can be found. It was found that the waves equilibrated at quite small amplitude, in contrast with the results of much more recent work by Ganser and Drew (1990), who integrated the one-dimensional equations numerically and found that the travelling waves developed into large amplitude structures similar to those predicted by Fanucci et al. (1979). In the linear approximation the equations of motion can be combined to give a single differential equation for the void fraction, as was shown quite early, by Jackson (1963) for example. Needham and Merkin (1983) solved an initial value problem for this equation, and also formulated an approximate differential equation for the void fraction which included nonlinear effects. (Exact elimination of all the other variables to yield a single equation for the void fraction is not possible with the full nonlinear equations.) Solution of an initial value problem for this approximate equation gave some indication of the development of sharp fronts, but the results were disappointingly inconclusive, and the approximations made in deriving the equation for the void fraction are controversial.

Quite recently Sasa and Hayakawa (1992) have reported solutions of the full one-dimensional equations in the form of solitary waves, but the significance of these is not yet clear.

To summarize the above work, there is an indication that initially sinusoidal, one-dimensional disturbances rising through a fluidized bed develop eventually into periodic structures for which the void fraction has a form of the type sketched in Figure 6, when plotted against height.

A completely different approach to the problem has recently been taken by Danckworth and Sundaresan (1991) and by Göz (1991, 1993), who sought fully-developed, periodic, one-dimensional profiles, propagating vertically at constant speed, that would satisfy the full nonlinear equations. By viewing these profiles from a frame of reference moving with them the equations of motion can be reduced to ordinary differential equations, and periodic solutions of these form limit cycles, which are well known in nonlinear mechanics. Bifurcation analysis then shows how solutions of this sort can branch away from the trivial solution representing the uniform bed as parameters in the equations (for example, the flow of fluidizing fluid) are varied. It was found that there is a domain of stability for the uniform bed, namely that predicted by linear stability theory, but that there are also branches representing solutions of the general sort sketched in Figure 6. An important conclusion of this work is that there is no qualitative difference between the predicted branching patterns and the nature of the solutions, between typical gas and liquid fluidized beds. Thus, the answer to the "bubbling versus non-bubbling" puzzle can never be found in one-dimensional solutions.

Secondary, two-dimensional instabilities of growing one-dimensional waves in a liquid fluidized bed have been described by El-Kaissy and Homsy (1976) and Didwania and Homsy (1981a), and a tentative theoretical explanation of the observations was provided by Didwania and Homsy (1982). Didwania and Homsy (1981b) have also examined a Rayleigh-Taylor type of instability of a density discontinuity introduced deliberately by suddenly increasing the flow of fluidizing medium. More recently Batchelor (1991) has suggested that several stages may be identified in the growth of a fully developed bubble from a small disturbance. First the uniform bed is disturbed by the fastest growing instability, namely a one-dimensional rising wave, then this wave is distorted by a secondary, two-dimensional motion which buckles the wavefronts and initiates segregation into regions of higher and lower void fraction in the lateral direction. Toroidal vortices then develop in the motion of the gas, and particles are expelled centrifugally to give the characteristic voids.

NUMERICAL INTEGRATION OF THE FULL EQUATIONS OF MOTION

At the beginning of this paper I noted some initial success in direct numerical solution of the full, time-dependent equations of motion. First Pritchett et al. (1978) and later Gidaspow and co-workers (Ettehadia et al., 1984; Gidaspow et al., 1985) have generated solutions which show credible looking bubbles rising from points where gas is injected at the distributor of a gas fluidized bed. Recently Kuipers et al. (1991) reported a careful solution which followed the motion of a pocket of gas injected into the fluidized bed, and showed development into a structure which seems to show all the observed characteristic properties of bubbles. Thus, there is reason to suppose that equations of the commonly used form have solutions which are good representations of real bubbles. However, none of the above work shows how a bubble might develop spontaneously from a small voidage perturbation, nor does it explain why the disturbances develop quite differently in typical gas and liquid fluidized beds.

The development of a small disturbance in a gas fluidized bed has been followed by Hernandez and Jiminez (1991), who integrated the equations of motion numerically from initial conditions corresponding to a disturbance which is periodic in both the horizontal and vertical directions, and noted the development of a structure resembling a two-dimensional array of bubbles.

At Princeton we have developed a code for the direct numerical integration of equations of the type (5) - (11) above and have used it to explore the development of small disturbances in both gas and liquid fluidized beds, following their development well out beyond the range of the linear approximation. I will now sketch the results of some recent computations of this kind by my student Karl Anderson. The fastest growing disturbance in a uniform fluidized bed is a rising, one-dimensional wave, with horizontal wavefronts, and integration of a small, initially sinusoidal disturbance of this kind shows a growth rate and wave speed in agreement with the predictions of the linearized equations while the amplitude remains small. As the amplitude grows a distortion of the sinusoidal waveform occurs, leading to a fully developed plane wave of the shape sketched in Figure 6. When viewed from a frame of reference moving with the wave, this is a steady, periodic solution of the equations of motion, and its stability against small perturbations can once more be examined. It turns out that this solution is subject to two types of instability. If it is constrained to remain one-dimensional, but the periodicity constraint is relaxed, there is a mode which eventually leads to coalescence of adjacent waveforms, giving another periodic solution which propagates without change of form, again of the general shape shown in Figure 6, but with a longer wavelength and larger amplitude than the original solution. This is of relevance to the physical behavior of a bed in a narrow tube.

Of more interest, however, is what happens when the constraint of one-dimensionality is relaxed. Then the fastest growing perturbation is two-dimensional, leading to structure along the wavefronts. The growth of this type of perturbation can be followed beyond the linear range by direct integration of the equations of motion, and the initial stages are quite similar in both gas and liquid fluidized beds, so to this point there is still no clue to a resolution of the puzzle of the difference in observed behavior of these two types of bed. A difference manifests itself only in the last stages of growth, where it becomes quite spectacular and demonstrates a rather simple underlying physical cause. To illustrate this Figures 7 - 10 show pictures of the density distribution and streamlines of the fluid flow field, at an intermediate stage of development of the two-dimensional instability, for both gas and liquid fluidized beds. Figures 7 and 8 refer to a bed of 200μm diameter glass beads fluidized by air at atmospheric pressure, with a volume fraction of particles of 0.57 for the uniform, unperturbed bed. Figures 9 and 10, on the other hand, correspond to a bed of 1mm diameter glass beads fluidized by water, with a solids volume fraction of 0.43 for the uniform, unperturbed bed. In each case a compact, two-dimensional region of decreased particle concentration has developed, in the neighborhood of which there is a closed vortex in the fluid flow field. Though the particle streamlines have not been shown, their velocity is directed downward everywhere, in both cases. All the views are from a frame of reference moving at a velocity quite close to that of the rising disturbance, and the reader is reminded that the patterns shown are repeated periodically in both directions, generating an infinite, two-dimensional array.

Figure 11 shows a much later stage in the development of the disturbance for the gas fluidized bed. The region of reduced particle concentration is now much deeper, and has assumed the characteristic, spherical cap shape of a bubble, with a sharply defined jump in density at its base and a rather more gradual change on passing through its roof. The gas streamlines, shown in Figure 12, contain the characteristic vortex of the Davidson cloud, moving with the bubble, while the solids streamlines of Figure 13 merely diverge to divert the particles around the rising bubble. A later stage of development of the disturbance in the liquid fluidized bed is shown in Figures 14 - 16. Figure 14 shows that the region of reduced particle concentration has not deepened, as it did for the bed fluidized by gas, and the reason for this can be seen on examining the streamlines. Figure 15 reveals a circulating vortex for the fluid, similar to that seen in Figure 12 for the gas fluidized case, but the particle streamlines of Figure 16 are quite different from the corresponding streamlines in the gas fluidized bed. Now the particle motion, as well as the fluid motion, contains a vortex. As a consequence of the upward flow of particles in the center of the vortex, dense material from below the region of reduced density is swept upward, filling in this region before it is able to grow into a bubble. Examination of the concentration distributions shows a tongue of dense material moving upward into the region of lower density at the time the vortex motion first appears in the particle velocity.

Earlier work was unable to find this vital difference in behavior between the two systems because it appears so late in the development of the disturbance. However, it takes us back to ideas on bubble stability proposed almost thirty years ago by Davidson and Harrison. Indeed, after almost fifty years of work and speculation by numerous investigators it appears that this turns out to be one of those happy cases in which one can say that "everyone was right". The two questions to be answered are: "why are bubbles found in most gas fluidized beds?" and: "why are they not found in most liquid fluidized beds?" The theory of stability of uniform beds reveals the basic instability without which the process of bubble growth cannot begin, and the concept of two-dimensional, secondary instability of the developing one-dimensional waves is needed to rescue the disturbance from growing into a final form which is essentially the same for all systems, and is not a bubble. Together then, these account for the ability of bubbles to grow in gas fluidized beds. On the other hand, the early ideas on bubble stability provide a mechanism which, as we have shown, can eventually differentiate between the gas and liquid fluidized systems and inhibit the last stages of growth into bubbles for liquid fluidized beds. Whether this is the universal mechanism which distinguishes bubbling from non-bubbling behavior will not be known until many other cases have been investigated computationally. Currently there are indications that other mechanisms may also play a role.

SIMPLIFIED FORMS OF THE EQUATIONS OF MOTION

The equations of motion used by my group in the past have been have been (5) - (8) above, with the closures given by equations (9), (10) and (11): others have also used these equations, or quite minor variants on them, so we should examine them a little more closely.

First we should note that they do not really describe a fluidized bed. The ideal behavior of such a bed is summarized in Figure 17a, which sketches the pressure drop and the bed expansion as functions of the superficial velocity of the fluidizing fluid. The pressure drop increases in proportion to the fluid velocity and the bed remains unexpanded up to the "minimum fluidization velocity", and over this range the assembly of particles has solid-like properties, exhibiting a non-vanishing yield stress. Beyond minimum fluidization, on the other hand, the pressure drop remains constant and equal to the bed weight, while the bed expands as the fluid velocity is increased. In the present equations p_s and μ_s are taken to be smooth functions of ε, which increase without bound as ε decreases to a value corresponding to random close packing. Then it is easy to see that the equations predict pressure drop and bed expansion curves of the sort sketched in Figure 17b. In contrast to Figure 17a, there is no break in slope of the curves to identify a condition of minimum fluidization; instead the bed is predicted to behave as a fluid throughout, but with a viscosity and pressure which increase without bound on approaching random close packing. Since $|dp_s/d\varepsilon| \rightarrow \infty$ as ε decreases toward close packing, there is always some value of ε greater than this limit for which $|dp_s/d\varepsilon|$ becomes large enough to stabilize the uniform bed. Our picture is, therefore, that the bed of particles expands continuously as the fluid velocity is increased from zero, but it behaves throughout as a fluid with effective viscosity and pressure which decrease continuously from infinite values at close packing. Also, the uniform bed remains stable up to some critical value of the expansion, at which point there is a transition to instability.

Second we note that the equations of motion have the correct limiting form when $\varepsilon \rightarrow 1$. Then the continuity equation for the fluid phase becomes $\nabla.\mathbf{u} = 0$, corresponding to an incompressible fluid with no suspended particles, while the momentum equation reduces to

$$\rho_f \frac{D_f \mathbf{u}}{Dt} = \nabla.\mathbf{E}_f + \rho_f \mathbf{g} \tag{16}$$

with

$$\mathbf{E}_f = -p\mathbf{I} + \mu\,[\nabla\mathbf{u} + \nabla\mathbf{u}^T] \tag{17}$$

that is, the Navier-Stokes equation for an incompressible fluid. The particle phase momentum equation reduces to the equation for the motion of a very dilute assembly of particles through a fluid medium, except for the non-local Basset force, which is absent.

Since the density and viscosity of a gas near ambient conditions are both small, for gas fluidized systems it is tempting to set $\rho_f = \lambda = \mu = 0$ in the momentum equations, when the momentum equation for the fluid reduces to

$$\nabla p + \beta(\varepsilon)(\mathbf{u} - \mathbf{v}) = 0 \tag{18}$$

describing the percolation of gas through the assembly of particles. Using this, and continuing to equate ρ_f to zero, the particle momentum equation then reduces to

$$\rho_s(1-\varepsilon)\frac{D_s\mathbf{v}}{Dt} = - \nabla.(p_s + p) + \nabla.\{\mu_s[\nabla\mathbf{v} + \nabla\mathbf{v}^T]\} + \rho_s(1-\varepsilon)\mathbf{g} \tag{19}$$

This looks like the equation of motion of a fluid of variable density $\rho = \rho_s(1 - \varepsilon)$, with effective viscosity $\mu_s(\varepsilon)$ and effective pressure $p_s(\varepsilon) + p$. However, since p_s is determined constitutively as a function of ε, while p must be determined dynamically, two equations of continuity are still needed, and the problem is not completely analogous to that of determining the motion of a variable-density fluid. In the limit when $\varepsilon \rightarrow 1$ equation (18) then reduces to $\nabla p = 0$, showing that the pressure is uniform in regions free of particles, to this approximation.

A simplification closely analogous to the above can be found even if the density and viscosity of the fluid may not be negligible. By combining equations (7) and (8) the following equation may be found for the slip velocity $\mathbf{w} = \mathbf{u} - \mathbf{v}$:

$$\frac{D_f\mathbf{w}}{Dt} + \mathbf{w}.\nabla\mathbf{v} + \varepsilon\beta(\varepsilon)\left[\frac{1}{\rho_f\varepsilon} + \frac{1}{\rho_s(1-\varepsilon)}\right]\mathbf{w} = -\left[\frac{1}{\rho_f} - \frac{1}{\rho_s}\right]\nabla.\sigma_f + \frac{1}{\rho_s(1-\varepsilon)}\nabla.\sigma_s \tag{20}$$

We can now define a relaxation time $\tau(\varepsilon)$ by

$$\frac{1}{\tau(\varepsilon)} = \varepsilon\beta(\varepsilon)\left[\frac{1}{\rho_f\varepsilon} + \frac{1}{\rho_s(1-\varepsilon)}\right] \tag{21}$$

and scaling analysis then shows that the first two terms on the left hand side of equation (20) can be omitted, to a good approximation, when τ is much shorter than all the relevant time scales of the problem considered.

Further straightforward manipulations then reduce the two momentum equations to the following form

$$[\rho_s(1-\varepsilon)+\rho_f\varepsilon]\frac{D_s\mathbf{v}}{Dt} = \nabla.\mathbf{E} + [\rho_s(1-\varepsilon)+\rho_f\varepsilon]\mathbf{g} \tag{22}$$

and

$$\beta(\varepsilon)\mathbf{w} = \rho_f\left(\mathbf{g} - \frac{D_s\mathbf{v}}{Dt}\right) + \nabla.\mathbf{E}_f \tag{23}$$

where $\mathbf{E} = \mathbf{E}_f + \mathbf{E}_s$. (22) is seen to be identical with the equation of motion of a single fluid with the same effective bulk density as the suspension, and a stress tensor equal to the sum of those associated with the separate phases. Equation (23) simply describes the percolation of the fluid through the particle assembly under the influence of the stress gradient and the effective gravity force, as seen by an observer in the frame of reference of the moving particles.

Although (22) is formally the same as the momentum equation of a homogeneous fluid with velocity **v** and variable density $\rho = \varepsilon\rho_f + (1-\varepsilon)\rho_s$, both continuity equations are still needed because, as we have seen above, the fluid phase pressure must be determined dynamically.

The approximation of short relaxation time is a good one in many cases, as can be seen from Table 1, which gives values of τ for 200μm glass beads suspended in air, and for 1mm glass beads suspended in water. In particular, at any but the most extreme values of dilution, it is clear that the relaxation time can be neglected in gas fluidized systems of particles smaller than 200μm, and of typical mineral density, even for problems with quite rapid macroscopic time scales.

We should finally address the most difficult outstanding question, namely the proper forms for the two stress tensors and the fluid-particle interaction force in the momentum equations. Most attempts to solve the equations have so far been based on the simple closure assumptions embodied in equations (9) - (11) above. However, we can probably do better than these in the case of gas fluidized beds, where the interactions between particles are primarily by direct solid-solid contact. When the particle concentration is not too near the limit of random close packing these contacts are primarily binary collisions, and results from the kinetic theory of granular materials (see, for example, Lun et al., 1984) can be invoked to give $\mathbf{E}_s$. The expression obtained is formally the same as (10) above but p_s, λ_s and μ_s now depend on the "granular temperature", in addition to the void fraction. (Granular temperature is defined as two thirds of the mean kinetic energy per unit mass of fluctuations in the particle velocity about **v**.) An extra equation, in the form of a balance for the pseudo-thermal energy of the particle phase, is then needed to complete the set of equations. This type of closure has been used in recent numerical solutions of the equations by Gidaspow and co-workers. Because of the screening effect of the particles and the low viscosity of the gas, it is also a good approximation to neglect the dynamic contribution to $\mathbf{E}_f$, simply writing the gas phase stress tensor as $-p\mathbf{I}$, provided the particle concentration is not too small.

Though the equations last described appear to be have a fairly sound basis, it must be emphasized that they apply only to gas fluidized systems in which the particle concentration is nowhere very low, or very high. When there are very few particles, as in the interior of bubbles, it is no longer justified to neglect the dynamic part of the gas phase stress tensor compared with that of $\mathbf{E}_s$. If this approximation is retained, in the limit of vanishing particle concentration the particle momentum balance does not reduce to the Navier-Stokes equation, as it should. Instead it reduces to the Euler equation, if the inertia of the gas is retained, or simply to the condition of constant gas pressure if it is not. At the opposite extreme of high particle concentration each particle may be in contact simultaneously with several neighbors, with normal and tangential forces exerted between particles at their points of rolling or sliding contact. The representation of the particle phase stress as a result of binary collisions is then inadequate, and there is presently no comparably fundamental theory available. However, empirical stress closures have been proposed in soil mechanics, and these include a representation of the yield stress which, as we have seen, is absent from the closures considered so far. Last year Roland Clift suggested that this type of stress closure might be needed to give an acceptable account of important phenomena in dense fluidized beds. In this he may well be correct, and such a closure can be introduced, though at the cost of a considerable increase in complexity and, probably, in difficulty when attempting to generate numerical solutions.

I close, then, by acknowledging that the goal of deriving exactly correct equations describing the mean motion, from the equations of motion of individual particles and the interstitial fluid, may be unattainable now, and for an indeterminate time in the future. Nevertheless, I re-emphasize that, for practical purposes, the goal is a set of equations which is good enough to describe the behavior of a range of situations of interest, to acceptable engineering accuracy. Perhaps, then, our objective should be to seek how much can be left out, while still meeting this criterion, rather than attempting to include every conceivable term.

LITERATURE CITED

Anderson, T.B. PhD Thesis, University of Edinburgh, 1967.

Anderson, T.B. and Jackson, R. Ind. Eng. Chem. Fundam. **6**, 527 (1967).

Anderson, T.B. and Jackson, R. Ind. Eng. Chem. Fundam. **7**, 12 (1968).

Anderson, T.B. and Jackson, R. Ind. Eng. Chem. Fundam. **8**, 137 (1969).

Batchelor, G.K. J. Fluid Mech. **193**, 75 (1988)

Batchelor, G.K. Symposium on Mechanics of Fluidized Beds, Stanford University, 1991.

Danckworth, D.C. and Sundaresan, S. Symposium on Mechanics of Fluidized Beds, Stanford University, 1991.

Davidson, J.F. Trans. Inst. Chem. Eng. **39**, 230 (1961).

Didwania, A.K. and Homsy, G.M. Int. J. Multiphase Flow **7**, 563 (1981a).

Didwania, A.K. and Homsy, G.M. Ind. Eng. Chem. Fundam. **20**, 318 (1981b).

Didwania, A.K. and Homsy, G.M. J. Fluid Mech. **122**, 433 (1982).

El Kaissy, M.M. and Homsy, G.M. Int. J. Multiphase Flow **2**, 379 (1976).

Ettehadieh, B., Gidaspow, D. and Lyczkowski, R.W. AIChE Jl. **30**, 529 (1984).

Fanucci, J.B., Ness, N. and Yen, R-H J. Fluid Mech. **94**, 353 (1979).

Foscolo, P.V. and Gibilaro, L.G. Chem. Eng. Sci. **39**, 1667 (1984).

Ganser, G.H. and Drew, D.A. Int. J. Multiphase Flow **16**, 447 (1990).

Garg, S.K. and Pritchett, J.W. J. Appl. Phys. **46**, 4493 (1975).

Geldart, D. Powder Tech. **7**, 285 (1973).

Gidaspow, D., Syamlal, M. and Seo, Y.C. 10th Powder and Bulk Solids Conference, 1985.

Gidaspow, D. and Therdthianwong, A. Proc. International Conference on Circulating Fluidized Beds, 436 (1993).

Göz, M.F. Symposium on Mechanics of Fluidized Beds, Stanford University, 1991.

Göz, M.F. Physica D **65**, 319 (1993).

Hernandez, J.A. and Jiminez, J. In: "The Global Geometry of Turbulence", NATO ASI Ser. B **268**, ed. J. Jiminez, Plenum Press, 1991.

Hinze, J.O. Appl. Sci. Res. **A11**, 33 (1962).

Jackson, R. Proc. Inst. Chem. Eng. Symp. on Interaction between Fluids and Particles, p42 (1962).

Jackson, R. Trans. Inst. Chem. Eng. **41**, 13, 22 (1963a,b).

Kuipers, J.A.M., Prins, W. and van Swaaij, W.P.M. Symposium on Mechanics of Fluidized Beds, Stanford University, 1961.

Liu, J.T.C. Proc. Roy. Soc. **A380**, 229 (1982).

Liu, J.T.C. Proc. Roy. Soc. **A389**, 331 (1983).

Lun, C.K.K., Savage, S.B., Jeffrey, D.J. and Chepurniy, N. J.Fluid Mech. **140**, 223 (1984).

Murray, J.D. J. Fluid Mech. **21**, 465; **22**, 57 (1965a,b).

Mutsers, S.M.P. and Rietema, K. Powder Tech. **18**, 239, 249 (1977a,b).

Needham, D.J. and Merkin, J.H. J. Fluid Mech. **131**, 427 (1983).

Pigford, R.L. and Baron, T. Ind. Eng. Chem. Fundam. **4**, 81 (1965).

Pritchett, J.W., Blake, T.R. and Garg, S.K. AIChE Symp. Ser. **176**, no. 74, 134 (1978).

Richardson, J.F. and Zaki, W.N. Trans. Inst. Chem. Eng. **32**, 35 (1954).

Rietema, K. Chem. Eng. Sci. **28**, 1493 (1973).

Rietema, K. and Piepers, H.W. Chem. Eng. Sci. **45**, 1627 (1990).

Rietema, K. "The Dynamics of Fine Particles", Elsevier, 1991.

Rowe, P.N. and Partridge, B.A. Chem. Eng. Sci. **18**, 511 (1963).

Sasa. S. and Hayakawa, H. Europhys. Lett. **17**, 685 (1992).

Slis, P.L., Willemse, Th.W. and Kramers, H. Appl. Sci. Res. **8**, 209 (1959).

Tsinontides, S. and Jackson, R. J. Fluid Mech. **255**, 237 (1993).

van Deemter, J.J. and van der Laan, E.T. Appl. Sci. Res. **A10**, 102 (1961).

Verloop, J. and Heertjes, P.M. Chem. Eng. Sci. **25**, 825 (1970).

Wilhelm, R.H. and Kwauk, M. Chem. Eng. Progr. **44**, 201 (1948).

TABLE 1: Numerical values of relaxation time in two systems

Air Fluidized bed. 200μm glass beads.

ε	τ(msec)
1	270
0.99	14
0.9	0.90
0.8	0.24
0.7	0.08
0.6	0.03
0.5	0.01
0.4	0.002

Water Fluidized bed. 1mm glass beads.

ε	τ(msec)
1	122
0.9	70
0.8	35
0.7	17
0.6	7.7
0.5	3.0
0.4	1.0

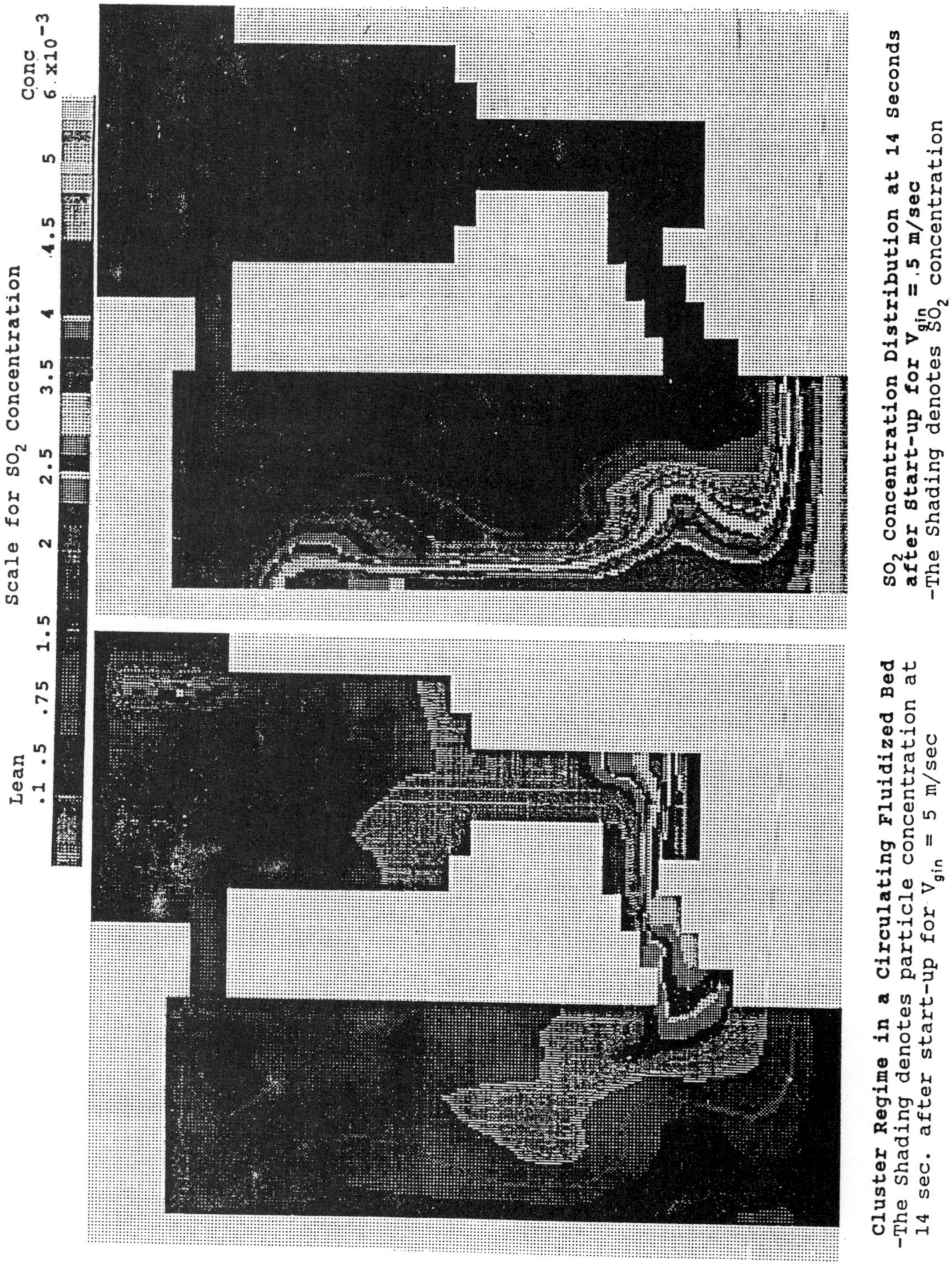

Figure 1. Simulation of a CFB SO_2 absorber. (After Gidaspow and Therdthianwong, 1993)

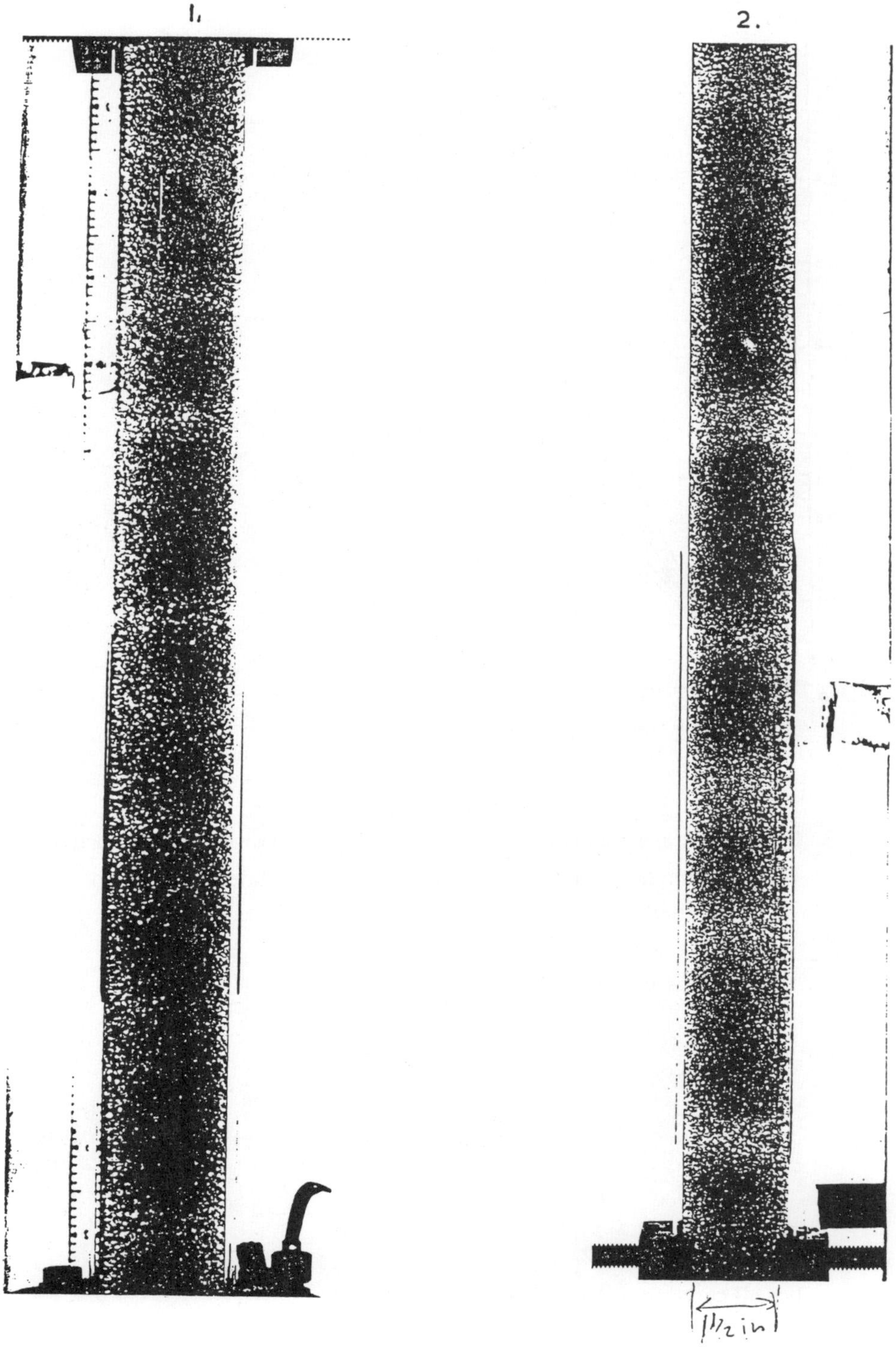

Figure 2. Instability waves in a 1.5in diameter water fluidized bed of 1mm glass beads. The right hand section of the diagram shows the part of the bed immediately above the section on the left, whose base is the distributor.

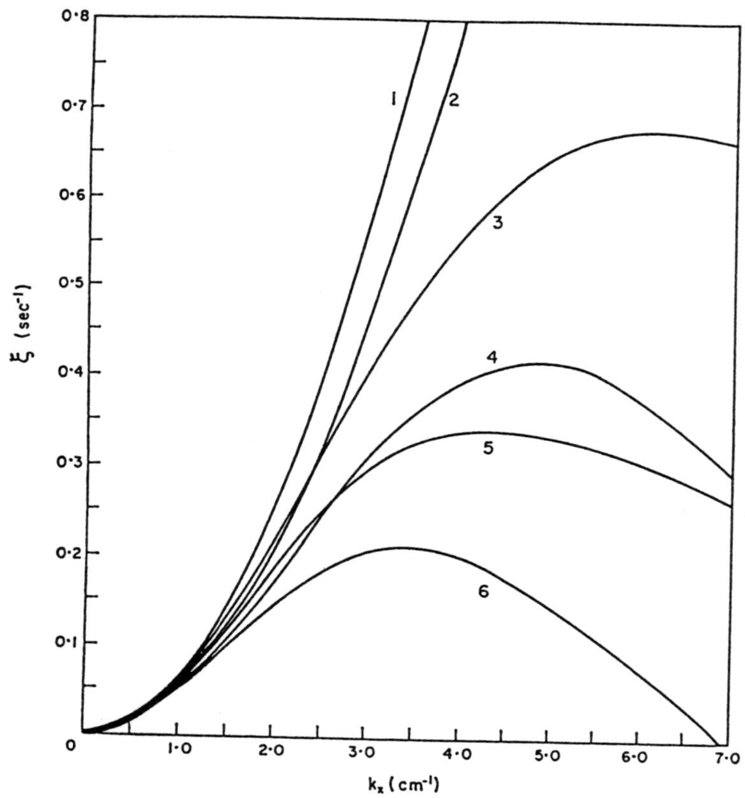

Figure 3. Growth rate as a function of wave number for small disturbances in a uniform bed of 1mm glass beads fluidized by water, as calculated by Anderson and Jackson, 1968.

Curve number	$\lambda_o{}^s + 4/3\mu_o{}^s$ poise	$\lvert dp^s/d\varepsilon \rvert_o$ dyn/cm^2
1	0	0
2	0	18
3	10	0
4	10	18
5	20	0
6	20	18

For all curves $\varepsilon_o = 0.46$

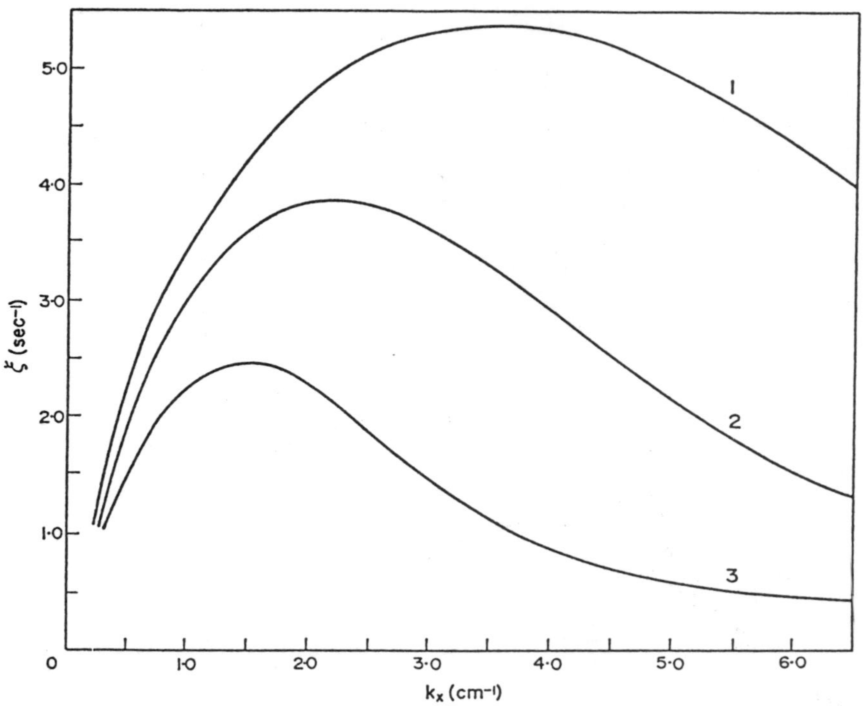

Figure 4. Growth rate as a function of wave number for small disturbances in a uniform bed of 1mm glass beads fluidized by air, as calculated by Anderson and Jackson, 1968.

Curve number	$\lambda_o^s + 4/3\mu_o^s$ poise
1	8
2	16
3	32

For all curves $\varepsilon_o = 0.42$; $|dp^s/d\varepsilon|_o = 20$ dyn/cm^2.

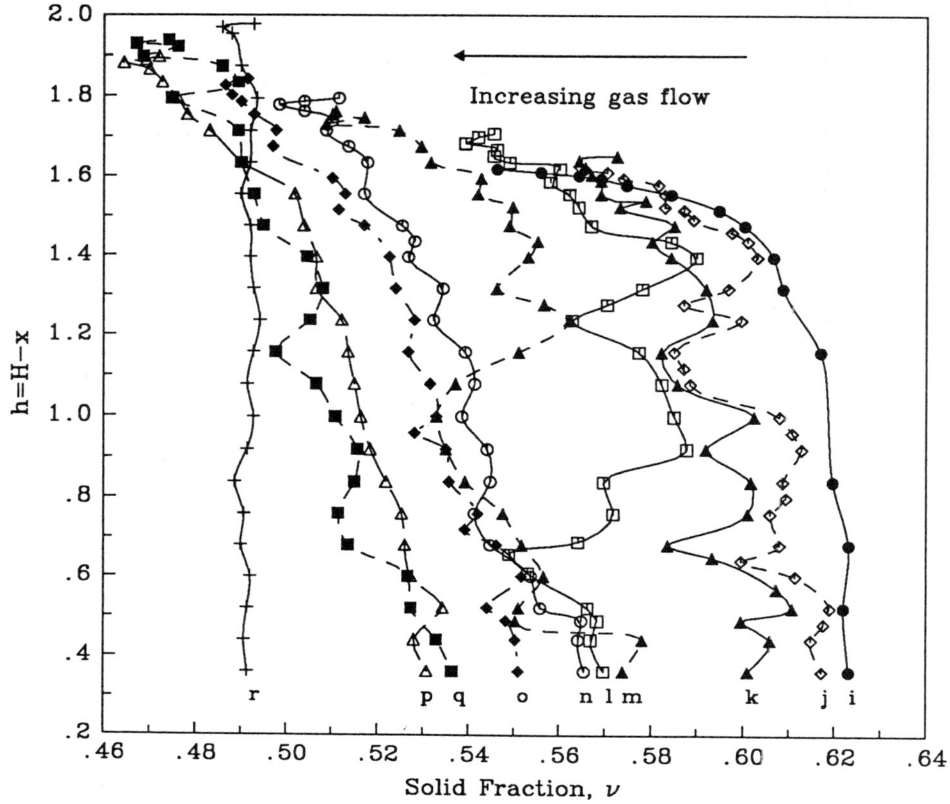

Figure 5. Axial profiles of solids fraction for expanded beds of FCC catalyst within the interval of "stable fluidization", as reported by Tsinontides and Jackson, 1993. The profile on the extreme right corresponds to zero gas flow, while that on the extreme left corresponds to a bubbling bed just beyond the point of initiation of bubbling. All measured values are time averages over at least ten minutes.

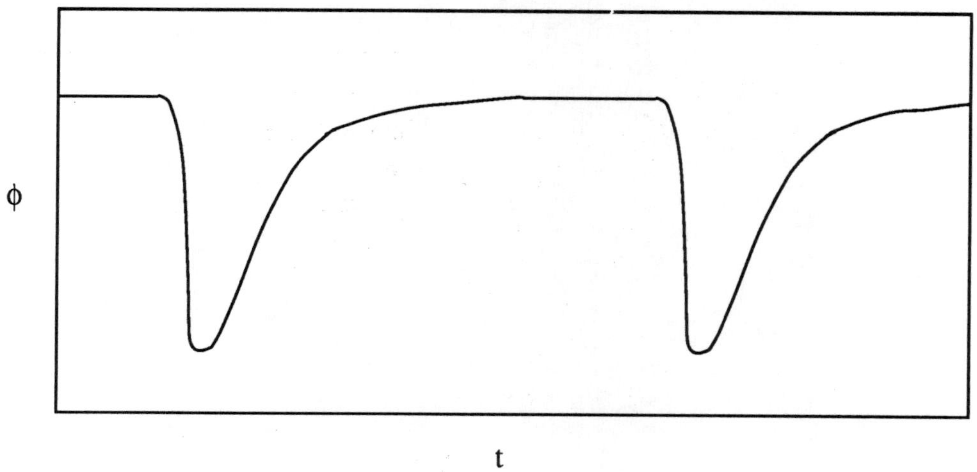

Figure 6. Sketch of the waveform of fully-developed one-dimensional waves rising through a fluidized bed.

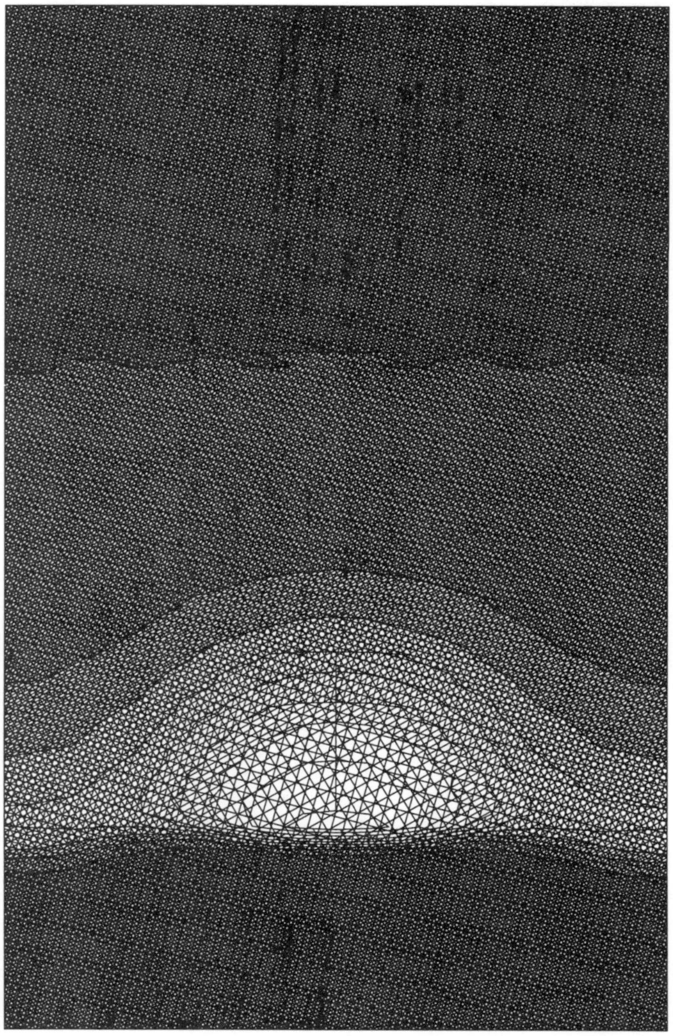

Figure 7. Predicted instantaneous density distribution in an air fluidized bed of 200μm glass beads, 0.26 sec. after a fully-developed one-dimensional wave is perturbed by a small two-dimensional increment proportional to the most unstable two-dimensional mode. Black represents close packed particles and white corresponds to particle-free gas.

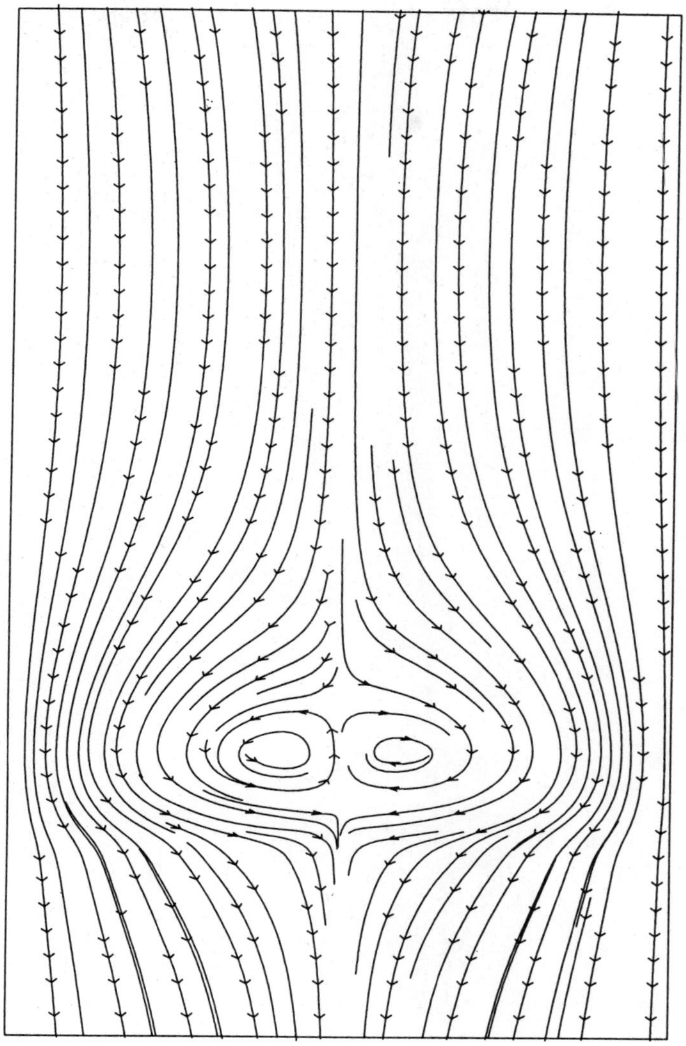

Figure 8. Predicted intantaneous streamlines for the gas velocity corresponding to Figure 7, illustrating the appearance of a vortex in the gas motion.

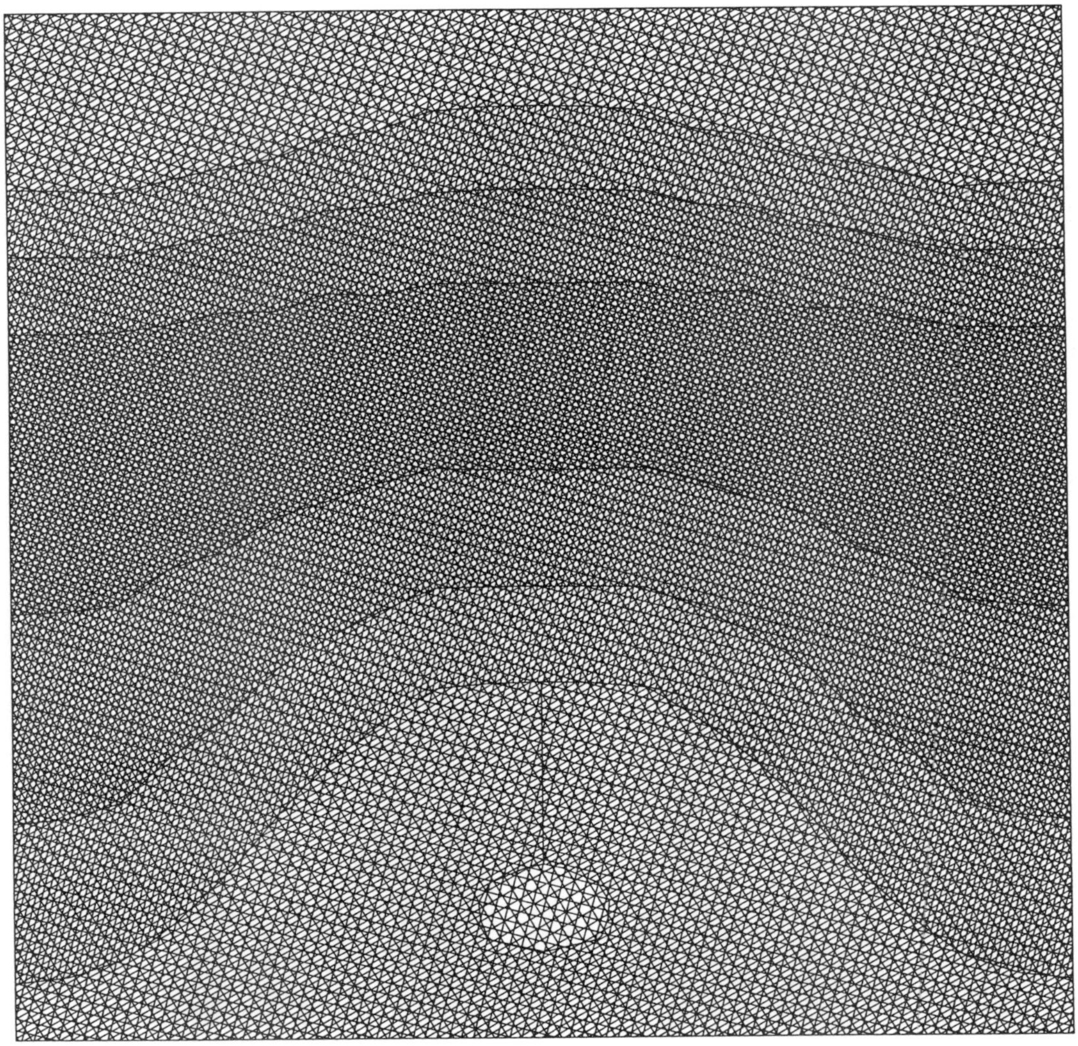

Figure 9. Predicted instantaneous density distribution in a water fluidized bed of 1mm glass beads, 0.63 sec. after a fully-developed one-dimensional wave is perturbed by a small two-dimensional increment proportional to the most unstable two-dimensional mode. Black represents close packed particles and white corresponds to particle-free gas.

Figure 10. Predicted instantaneous streamlines for the water velocity corresponding to Figure 9, illustrating the appearance of a vortex in the liquid motion.

Figure 11. As Figure 7 (air fluidized bed) but 0.52sec. after the introduction of the two-dimensional perturbation.

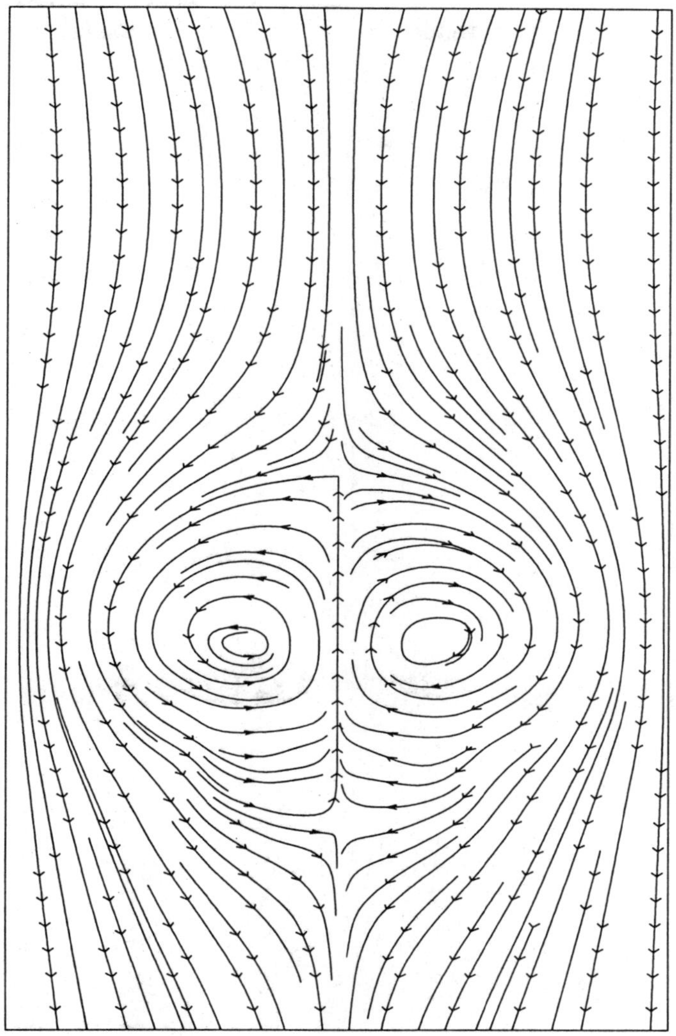

Figure 12. Predicted instantaneous streamlines of the gas velocity corresponding to Figure11.

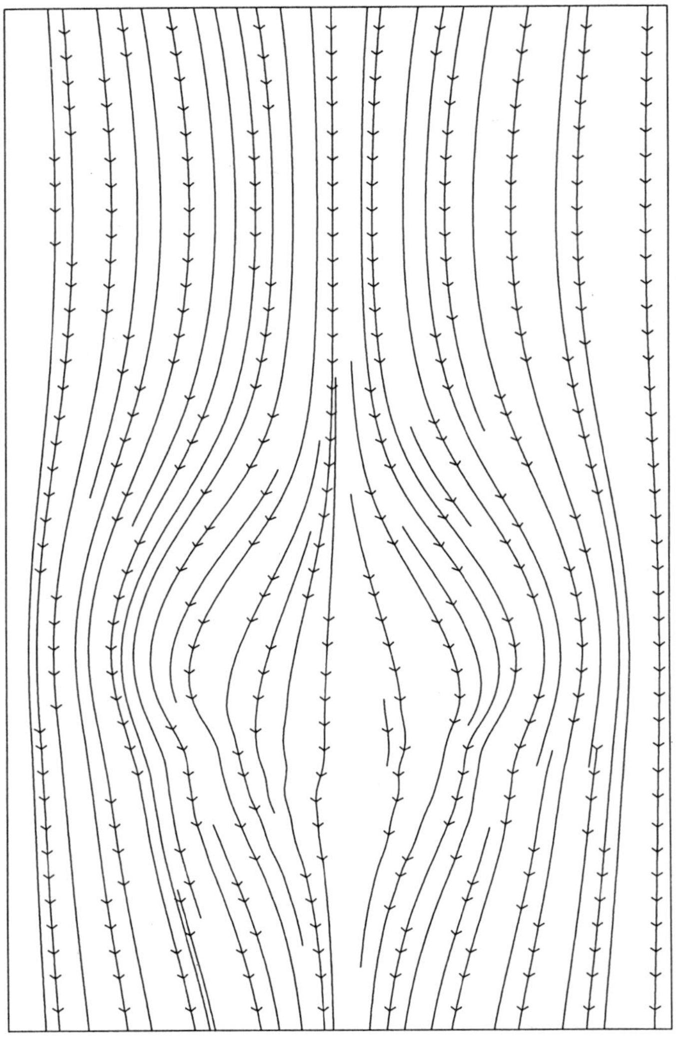

Figure 13. Predicted instantaneous streamlines of the particle velocity corresponding to Figure 11.

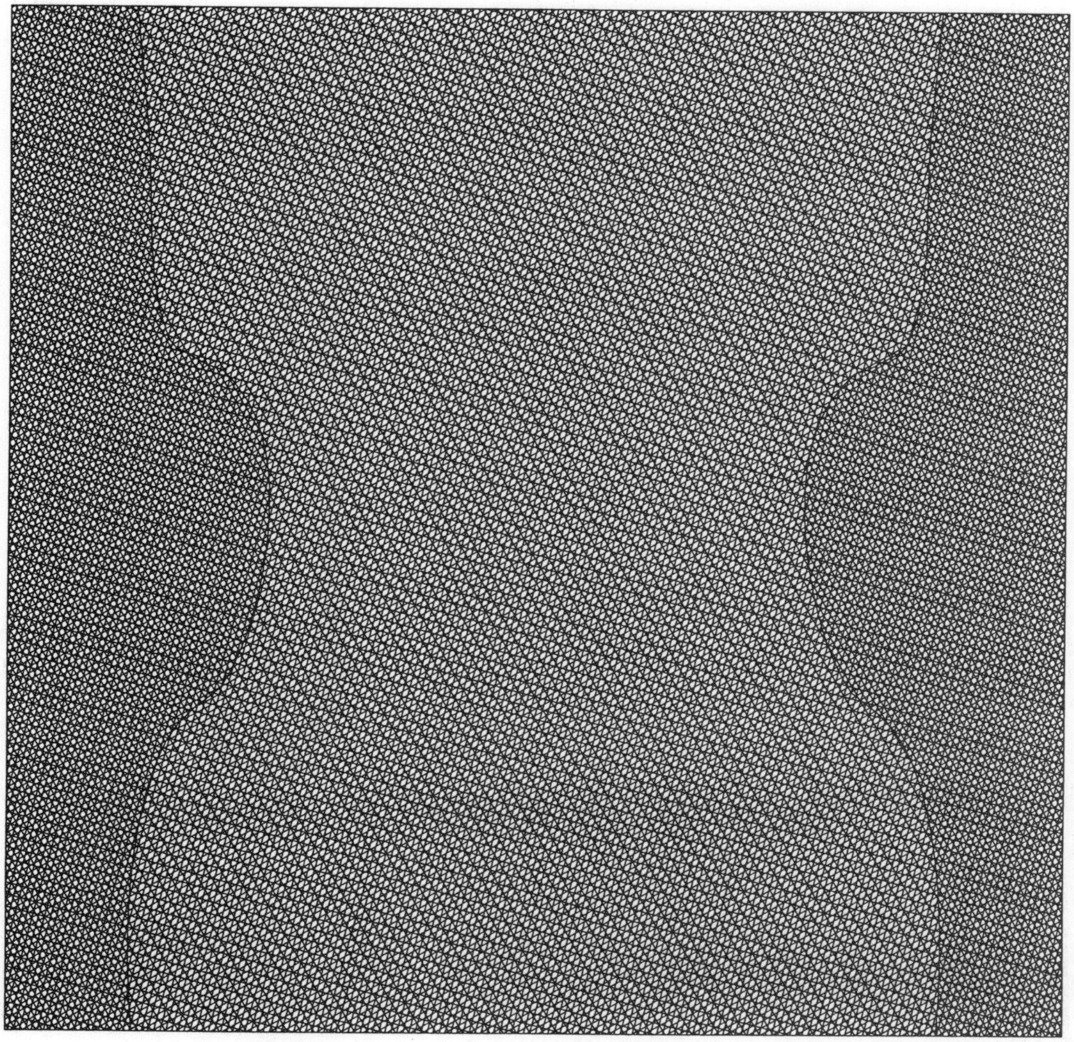

Figure 14. As Figure 9 (water fluidized bed) but 3.1 sec. after the introduction of the two-dimensional perturbation.

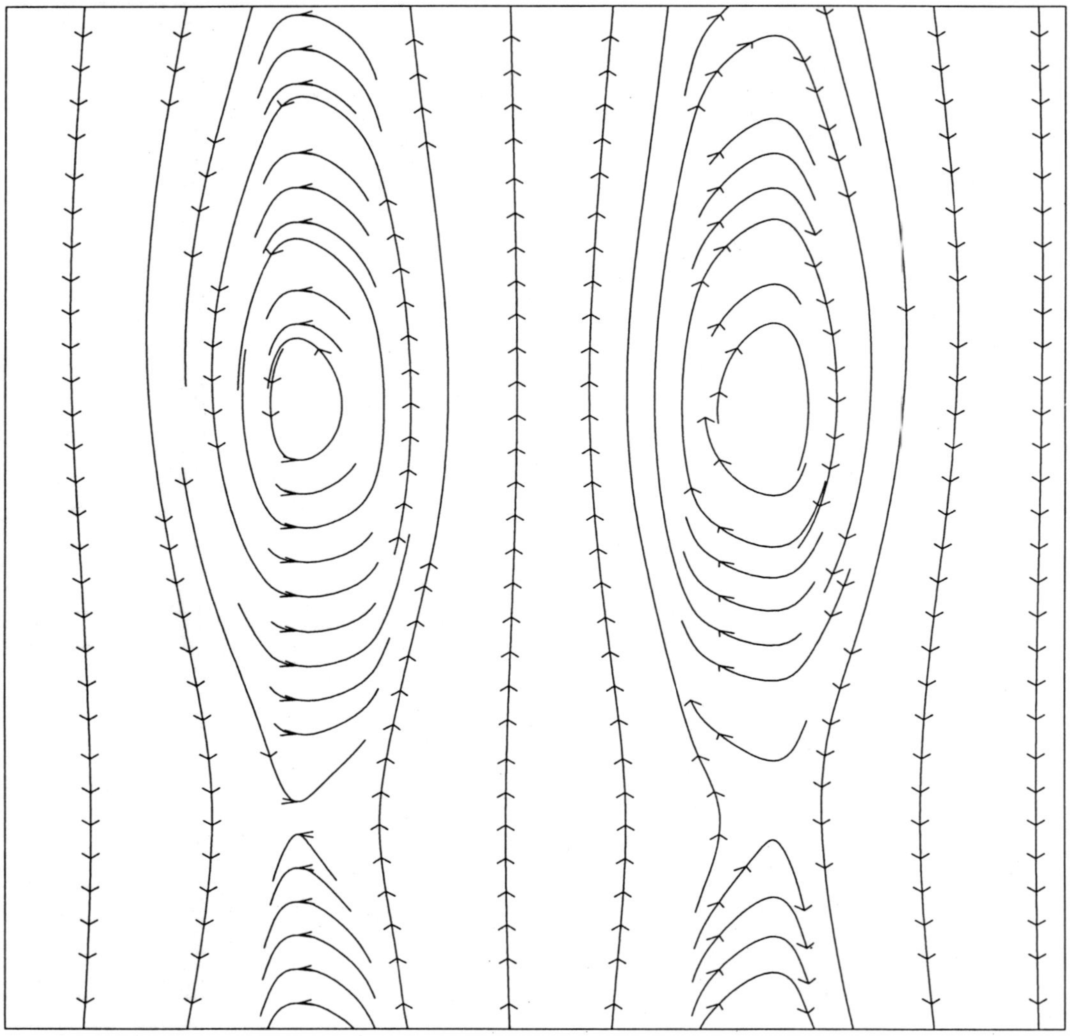

Figure 15. Predicted instantaneous streamlines of the water velocity corresponding to Figure 14.

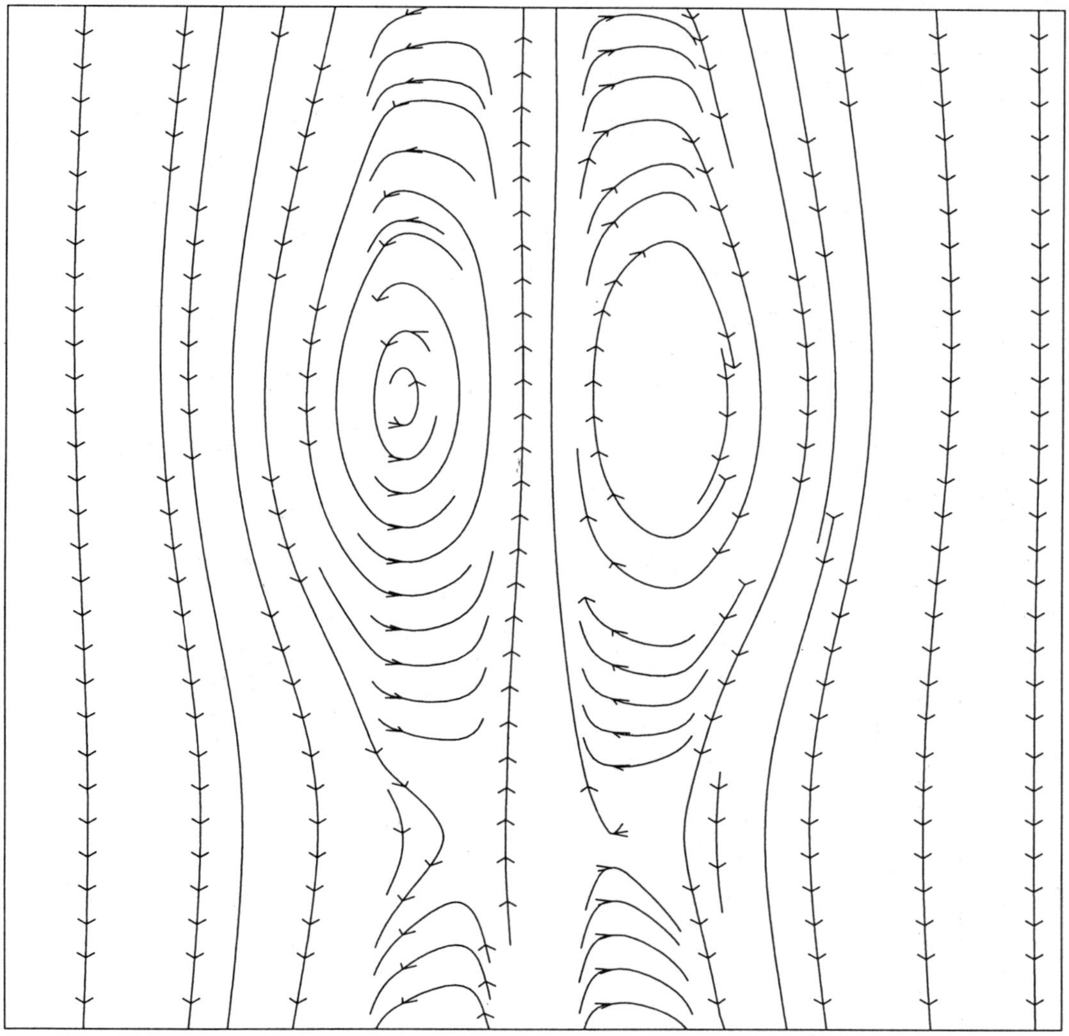

Figure 16. Predicted instantaneous streamlines of the particle velocity corresponding to Figure 14.

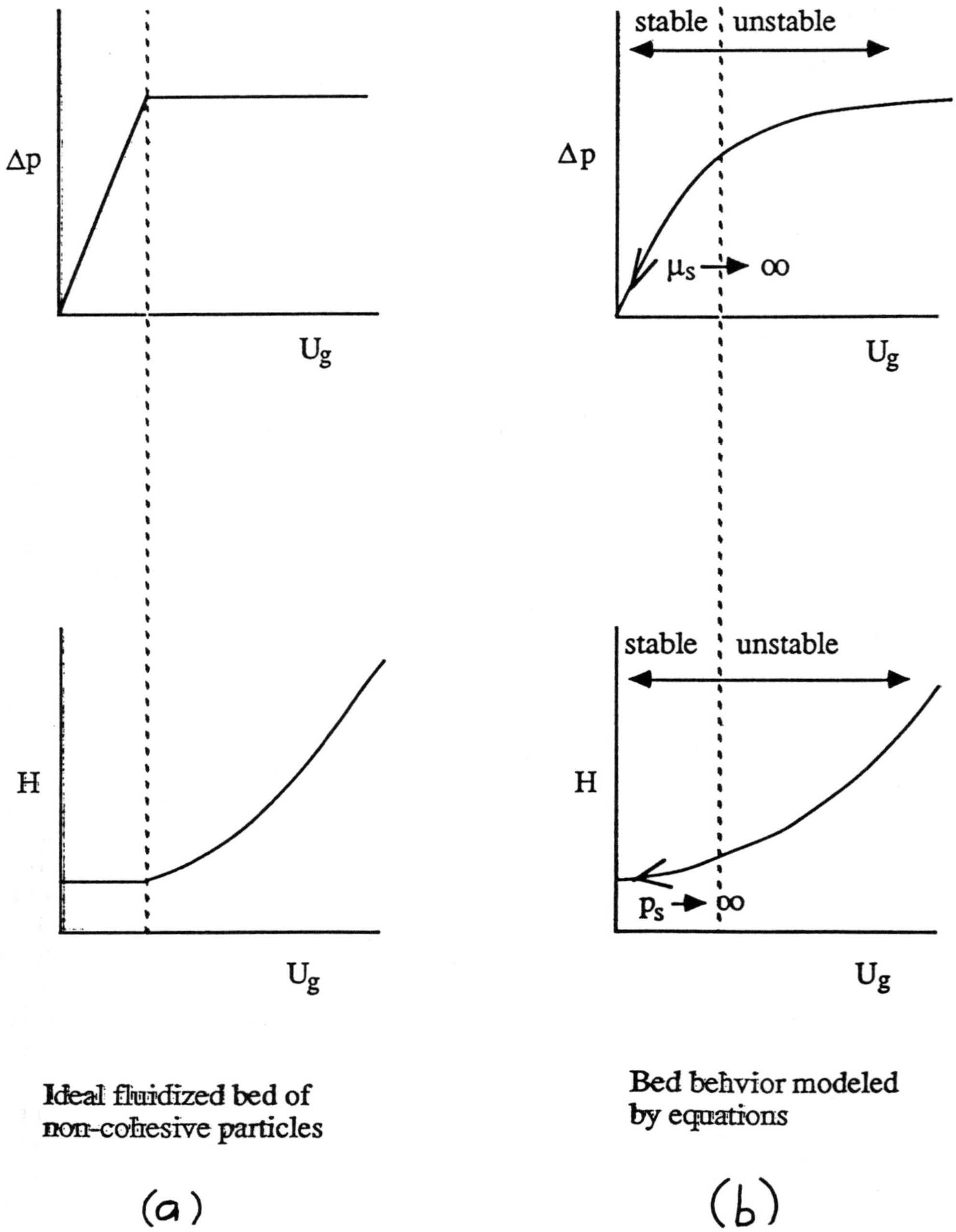

Figure 17. Bed pressure drop and bed expansion curves for an ideal fluidized bed, contrasted with the predictions of equations with Newtonian stress closures.
(a) ideal fluidized bed,
(b) bed with Newtonian stress closures.

Expansion Behavior of Gas Fluidized Beds in the Turbulent Regime

J. Werther and J. Wein
Technical University Hamburg-Harburg, Denickestr. 15, D-21071 Hamburg, Germany

The aim of the present work was to find a description for the axial distribution of solid particles in gas fluidized beds operated at high gas velocities. Measurements of local solids concentrations were carried out in beds of 0.5 x 0.5 m^2 cross-section and 0.6 m diameter. The beds were operated at gas velocities u between 0.3 and 4 m/s. At the higher gas velocities, between 50 and 70 wt.- percent of the bed material was entrainable according to the distributions of the particles' terminal velocities.

Contrary to other authors' findings in beds of smaller diameters, no significant changes in the fluidization characteristics were observed when the gas velocity was increased from bubbling to turbulent fluidization. Basically the same two-phase structure of the bed with gas voids in a dense suspension was always observed. At all fluidization conditions, the voids were rising at velocities which could be interpreted as the velocities of rising bubbles. An existing bubble coalescence model was found to be sufficient to describe the sizes of these voids. A calculation procedure is suggested for the prediction of the axial solids volume concentration profile which includes some newly developed correlations for the dense phase voidage and for the visible bubble flow. The calculated profiles of the solids concentration are shown to provide a good description of the measurements.

In the past decades the bubbling fluidized bed which was operated at velocities between 1 and 2 ft/s has received much attention. With the advent of the fluidized bed combustion process in the late 1970's and during the 80's the interest of researchers shifted to fluidized beds which were operated at higher gas velocities i.e. in the turbulent and in the fast fluidization regime. These high velocity systems are characterized by high rates of solid entrainment from the bed, the clearly discernible bed surface is replaced by a diffuse transition zone between the bed and the freeboard, and in the freeboard an exponential decay of the solids concentration with height is observed. A knowledge of the bed expansion and of the axial distribution of solids in bed and freeboard is essential for the modeling of fluidized bed combustors.

In spite of the practical relevance of the subject there are comparatively few papers available in the open literature which deal with the experimental investigation and modeling of fluid mechanics in the turbulent fluidization regime on the pilot or large scale. Many authors have instead dealt with pressure differential measurements in small diameter beds and have deduced characteristic parameters for the transition from the bubbling to the turbulent fluidization regime (e.g. [1]). Recently, Leckner and his co-workers [2-4] have studied the bottom zone of the circulating fluidized bed in a cold model and in the 12 MW circulating fluidized bed combustor at Chalmers University of Technology. They concluded that the bottom dense bed of the CFB had all charactistics of a bubbling bed.

In the present paper the results of an experimental investigation of fluid-mechanical characteristics in two pilot-scale cold model fluidized beds are presented together with the derivation of a calculation procedure for the axial solids volume concentration profile.

EXPERIMENTAL

Fluidized bed installations

Two fluidized bed units were used in the present work. One bed had a square cross-section

Technical University Hamburg-Harburg
Denickestr. 15, D-21071 Hamburg

with a side length of 0.5 m and a total height of 4 m. The second unit had a circular cross-section with 0.6 m diameter and a total height of 4.5 m in the beginning of the experiments which was later on extended to 6 m. Air as the fluidizing gas was supplied by two roots blowers (Aerzener Maschinenfabrik, GMa 13.8 and GMb 16.12). A schematic view of the second unit is shown in Figure 1. Figure 2 presents the design of the bubble cap distributor. Since it was difficult to provide an adequate distributor pressure drop over the whole range of operating gas velocities between 0.3 and 4 m/s the bubble caps were fitted with exchangeable caps at their lower ends. Several caps with different bore diameters were used in different ranges of fluidizing velocities. Entrained solids which are collected in two primary and one secondary cyclones are through 50 mm i.d. tubes returned directly into the bed. Purge air was sometimes used to initiate solids flow in the return lines. The location of the solids inlets relative to the gas distributor is shown in Figure 2. Both fluidizing columns are equipped with measuring ports for the introduction of various probes. The vertical spacing between the ports is 0.1 m in the bottom region and 0.5 m in the upper part of the column.

Bed material and operating conditions

A quartz sand fraction and bed ash from a German circulating fluidized bed combustor were used as bed materials. Some physical properties are given in Table 1. Figure 3 presents the particle size distributions together with the distributions of the single particles' terminal velocities calculated with the assumption of spherical particles. The comparison with the range of operating gas velocities indicates that at the higher gas velocities between 50 and 70 wt.-% of the bed particles are entrainable according to their terminal velocities.

Measuring techniques

Pressure differences were measured over height intervals Δh with pressure transducers. The pressure signals were recorded and from the time-average pressure drop Δp the solids volume concentration $c_{v,\Delta p}$ was calculated according to $c_{v,\Delta p} = \Delta p/(\Delta h\, \rho_s\, g)$ assuming that hydrostatics is the only contribution to Δp. $c_{v,\Delta p}$ averages over the bed's cross-sectional area and over the corresponding height interval Δh.

The attenuation of γ-rays from a Cs-137 source measured by a scintillation counter was used as a second method to determine a solids volume concentration. The resulting $c_{v,\gamma}$ is averaged along the path of the γ-rays through the bed.

Local solids volume concentrations were measured with capacitance probes of the needle type [5]: the protruding needle forms one pole, while the enclosing metal tube forms the other pole of the capacitor. Figure 4 illustrates the design of the probe which is water-cooled in order to stabilize the electric circuitry. The capacitor is part of a high frequency oscillator circuit. A change in the capacitance will change the frequency of the oscillator circuit which in turn will be converted to a corresponding change of a DC signal by a reactance converter (Dantec, 56 EO1). In the past the needle probe has often been used for measuring bubble sizes and velocities which is comparatively simple since it requires the distinction between bubble and suspension phases only whereas the determination of the local solids volume concentration requires a calibration of the probe.

Bakker and Heertjes [6] have shown that the relationship between the dielectric constant of the gas/solid mixture, D_m, and the solids volume concentration c_v may be described by a relationship which has been obtained by Wiener [7] for an isotropic arrangement of spheres,

$$\frac{D_m - D_f}{D_m + 2\,D_f} = c_v \, \frac{D_s - D_f}{D_s + 2\,D_f} \qquad (1)$$

where D_f and D_s are the relative dielectric constants of fluid and solid, respectively ($D_f = 1$ for air). The dielectric constant D_s was determined here by filling the annulus of a cylinder capacitor (100 mm outer dia., 88 mm inner dia., 257 mm long) with a fixed bed of the solid particles ($c_{v,fb} = 0.6$). The ratio of the capacitances of full and empty annulus, respectively, is identical to the relative dielectric constant of the fixed bed mixture $D_{m,fb}$ which may be used to calculate D_s from equ. (1) with $c_{v,fb} = 0.6$ and $D_f = 1$.

If the capacitance probe is exposed to a change in the local relative dielectric constants,

$$\Delta D = D_m(c_v) - D_f$$

this will cause a (small) change in the capacitance (≤ 0.5 pF) which is converted to a proportional change in the output voltageof the Dantec system,

$$\Delta U = U(c_v) - U_o$$

where U_o is the voltage at $c_v = 0$ which corresponds to $D_m = D_f$. The maximum change in the mixture dielectric constant is

$$\Delta D_{max} = D_{m,fb} - D_f$$

which corresponds to

$$\Delta U_{max} = U(c_{v,fb}) - U_o.$$

It follows

$$\frac{U(c_v) - U_0}{U(c_{v,fb}) - U_0} = \frac{D_m(c_v) - D_f}{D_{m,fb} - D_f} \qquad (2)$$

as the calibration function of the capacitance probe.

The probe is calibrated outside the fluidized bed by once recording its output signal in air ($c_v = 0$, $U = U_o$) and afterwards dipping the probe into a fixed bed of sand or ash particles ($c_v = c_{v,fb}$, $U = U(c_{v,fb})$).

Due to slight temperature variations during operation of the fluidized bed some drift of the probe signal is unavoidable. This may, however, be compensated by an in-situ recalibration where only one reference point is needed since the difference ($U(c_{v,fb})$ - U_o) is not affected by the drift. We take either the fixed bed point (during measurements at low gas velocities or in bubble-free regions of the bed) or the $c_v = 0$ point (with measurements at locations with high bubble concentrations or in the freeboard) obtained from amplitude probability density distributions of the probe signal.

Two optical fiber probes aligned vertically above each other of the single fiber reflexion type [8] were used to determine bubble rise velocities from cross-correlation measurements [5].

RESULTS AND DISCUSSION

A comparison of solids volume concentrations obtained by different methods

Figure 5 gives a comparison of solids volume concentrations calculated from pressure drop measurements with those obtained from the capacitance probe. The measurements were carried out in the 0.5 x 0.5 m^2 square bed with sand as a bed material. The bed height at minimum fluidization was 0.64 m. The pressure difference was measured between 0.34 m and 0.44 m above the distributor plate which corresponds to distances of 0.326 m and 0.426 m from the lower edge of the bubble caps. The average solids volume concentration obtained from the Δp measurement is seen to decrease slightly with increasing gas velocity from the fixed bed value of 0.6 to about 0.45 at a gas velocity of 3 m/s. The capacitance measurements were carried out at a level of 0.34 m above the distributor plate. Measurements at different distances from the wall yielded the lateral c_v-profile the integration of which over the bed's cross-sectional area yielded the average c_v values plotted in Figure 5. The comparison of the two methods shows that the solids concentration values obtained from the Δp method are always higher than those calculated from the capacitance measurements.

Since it was not clear which of the two methods might be more reliable it was decided to add a third independent method for the determination of the solids volume concentration. Therefore γ-ray attenuation measurements were carried out again in the 0.5 x 0.5 m^2 unit with sand as the bed material. The γ-ray source

(Cs 137) was mounted at the outside wall of the bed. The scintillation counter was placed outside the opposite wall. Measurements were made at different levels above the distributor. The resulting average solids volume concentrations are plotted in Figure 6 as a function of height above the distributor.

Since $c_{v\gamma}$ is averaged along the path of the γ-ray the local capacitance probe measurements were also averaged along a traverse through the bed. The comparison between the two methods was made for two different gas velocities. The lower one (u = 0.5 m/s) is typical of the bubbling bed regime whereas the higher one (u = 2.7 m/s) is representative of the turbulent regime. In both cases an excellent agreement between the two methods is obtained. Since the γ-method seems to be without systematic errors it may be concluded that the capacitance probe is a reliable tool for measuring local solids volume concentrations.

As a further conclusion it may be stated with reference to Figure 5 that the pressure drop method obviously yields too high values in the present case. The reason may lie in the acceleration of solid material in the zone close to the distributor which increases the pressure drop above the value that is necessary to keep the particles suspended and thus leads to the observed overestimation of the pressure drop.

Evaluation of capacitance probe signals

Figure 7 shows the time series of the probe signals obtained at two different gas velocities in the 0.6 m diameter bed. In Figure 8 the corresponding amplitude probability distributions are shown in the form of probability distributions of the local instantaneous solids volume concentrations, $p(c_v)$. At the lower velocity we observe the characteristics of a typical bubbling fluidized bed. The bubble pulses are clearly seen in the signal sequence. At the higher gas velocity the predominance of lower concentration values and the much lower time average of c_v may be taken as indications of the turbulent fluidization regime. However, the general two-phase structure of the signal remains much the same as is obvious from the comparison of the two probability distributions $p(c_v)$. In both cases we have a distribution with two peaks. The peak at lower concentrations is representative of the lean or bubble phase whereas the peak at the concentration range between 0.4 and 0.5 is characterizing the dense or suspension phase. With increasing gas velocity the lean phase portion of the concentration distribution grows at the expense of the dense phase part.

Figure 9 shows radial porosity distributions measured at different heights above the distributor plate for the same two velocities. At the lowest level (h= 0.14 m) we observe an annular zone of increased voidage in the bubbling bed which shifts with increasing height above the distributor to the vessel center. This voidage distribution which corresponds to an annular bubble flow pattern is typical of the bubbling bed and has been observed by a number of authors (e.g.[9,10]). In the freeboard, we see comparatively flat voidage profiles with slightly decreasing voidage close to the wall. The cross-sectional average voidage increases with increasing height above the distributor. In the turbulent region similar profiles $\varepsilon(r)$ are observed at all heights with the voidage always decreasing towards the wall.

Figure 10 compares the axial profiles of the cross-sectional average solids volume concentration for the two velocities. The bubbling bed is characterized by a solids concentration between 0.4 and 0.45 in the bed region with a steep decrease at the bed surface. On the other hand, the turbulent bed is characterized by lower solids concentrations around 0.25 in the bed region and a very smooth decay of the solids volume concentration in the freeboard. The splash region is seen to extend in this case over a height of roughly 2 m.

The c_v value corresponding to the dense phase peak in the $p(c_v)$ distribution (cf. Figure 8) is taken as a measure for characterizing the solids volume concentration of the dense phase c_{vd} or the dense phase voidage ε_d, respectively. Measurements of this quantity have been obtained for three different heights above the distributor (h = 0.14, 0.34, 0.54 m) at different distances r from the vessel centre. Since no clear influence of r on ε_d was observed the individual values ε_d (h,r) were in the first step averaged over the cross-sectional area of the bed. Since the dependence of the resulting ε_d(h) values on h was found insignificant the averages over h are plotted as a function of ($u - u_{mf}$) in Figure 11. For each value of ε_d the 95 % confidence interval is indicated. For both bed materials a significant increase of ε_d with increasing gas velocity is observed. This tendency is in agreement with other authors' findings [11,12].

Modeling of the axial profile of the solids volume concentration

The present modeling approach distinguishes between the bed region and the freeboard. The bed region model which will be treated first is based on the following assumptions:

(1) The bed region is composed of two phases, a dense phase and a void phase.
(2) The void phase consists of bubbles which are assumed to be solids-free.
(3) All bed properties are considered to be independent of the radius, i.e. the model considers cross-sectional average characteristics only.

The assumption (2) will be examined in detail in section 3.4 of this paper. With the foregoing assumptions the solids volume concentration c_v at a height h above the distributor is given by

$$c_v = (1 - \varepsilon_b)\, c_{vd} \tag{3}$$

where ε_b denotes the bubble volume fraction.

The measurements have shown that c_{vd} is independent of height h. Instead a dependence of c_{vd} on gas velocity and bed material properties was observed. In order to obtain a quantitative relationship the ratio c_{vd}/c_{vmf} was correlated with a particle Reynolds number based on the excess gas velocity ($u - u_{mf}$),

$$Re_p = (u - u_{mf})\, d_{ps}/\nu, \tag{4}$$

and with the Archimedes number

$$Ar = g\, \frac{d_{ps}^{\,3}}{\nu^2}\, \frac{\rho_s - \rho_f}{\rho_f} \tag{5}$$

The resulting correlation is

$$\frac{c_{vd}}{c_{vmf}} = 1 - 0.14\, Re_p^{0.4}\, Ar^{-0.13} \tag{6}$$

which contains as a limiting case $c_{vd} = c_{vmf}$ for $u = u_{mf}$ and which is seen to give a good representation of the measured data in Figure 12. It should be noted, however, that Equation (6) is based on cold model experiments with two bed materials only. Further work is certainly needed to investigate other influences e.g. the temperature effect on c_{vd}.

The bubble volume fraction ε_b is calculated from

$$\varepsilon_b = \dot{V}_b / u_b \tag{7}$$

where $\dot{V}_b$ is the visible bubble flow rate based on unit bed area and u_b is the (average) rise velocity of the bubbles at the height h above the distributor. $\dot{V}_b$ may be related to the excess gas velocity ($u - u_{mf}$) by

$$\dot{V}_b = \varphi\,(u - u_{mf}) \tag{8}$$

According to previous work of the authors' group (e.g.[13]) the bubble rise velocity may be calculated from

$$u_b = \dot{V}_b + 0.71\, \vartheta \sqrt{g\, d_v} \tag{9}$$

where the parameter ϑ is describing the influence of the bed diameter d_t,

$$\vartheta = \begin{cases} 0.63 & d_t < 0.1 \text{ m} \\ 2.0\sqrt{d_t} & 0.1 \text{ m} \le d_t \le 1.0 \text{ m} \\ 2.0 & d_t > 1 \text{ m} \end{cases} \tag{10}$$

for a Geldart B powder. The bubble volume equivalent diameter d_v in Equation (9) may be calculated from the differential equation

$$\frac{d(d_v)}{dh} = \left(\frac{2\,\varepsilon_b}{9\,\pi}\right)^{1/3} \qquad (11)$$

for a Geldart B powder with the boundary conditions at $h = h_o$:

$$d_{v0} = 1.3 \left(\frac{\dot{V}_0^{\,2}}{g}\right)^{0.2} \qquad (12)$$

for a technical gas distributor. In the present case of a bubble cap distributor with 67 bubble caps/m^2 bed area $\dot{V}_0$ was assumed to be equal to the flow rate through an individual nozzle i.e. each bubble cap was considered to act as a single bubble generating orifice.

Experimental values of the bubble volume fraction ε_b have been obtained from the capacitance probe signal u(t) or $c_v(t)$, respectively, as the ratio of time during which $c_v < 0.15$ was observed to the total measuring time. This cut-off level has been selected such that the c_v-value calculated from Equation (3) was in agreement with the measured time-average c_v. If ε_b is determined from the measurements it is possible to calculate values of the parameter φ by iterative use of eqns. (7) - (12) with a starting value of $\varphi = 0.76$ which has previously been obtained for sand with $d_{ps} = 103$ μm and a technical gas distributor [14].

The resulting values of φ are plotted in Figure 13 as a function of height h above the distributor. In the case of sand no significant influence of height is observed. With increasing gas velocity φ decreases from about 0.4 to about 0.2 at the velocity of 1.58 m/s but increases again with further increasing gas velocity. In the case of ash a decrease of φ with both height and gas velocity is observed when the gas velocity is increased from 0.38 m/s to 1.04 m/s. For gas velocities between 1.44 m/s and 2.05 m/s φ is fairly independent of u and h, respectively.

Since the whole picture is difficult to explain at the moment the influences of u and h are neglected as a first approximation. The arithmetic average over all measurements yields $\varphi = 0.37$ for the sand and $\varphi = 0.50$ for the ash, respectively. Taking additional results from previous work [14, 15] into account the parameter φ has been plotted in Figure 14 against the Archimedes number. It follows as an empirical correlation

$$\varphi = 1.45\ Ar^{-0.18} \qquad 10^2 < Ar < 10^4 \qquad (13)$$

With the set of equations presented above the axial solids concentration profile in the bed region may now be predicted. For the description of $c_v(h)$ in the freeboard region the simple exponential decay function [16,17],

$$c_v(h - h_B) - c_{v\infty} = (c_v(h_B) - c_{v\infty})\, e^{-a\,(h-h_B)} \qquad (14)$$

is adopted where h_B denotes the expanded bed height and $c_{v\infty}$ is the limiting concentration for $h \rightarrow \infty$. The decay constant a may be obtained from the $c_v(h)$ profiles measured in the turbulent regime which are characterized by a smooth transition from the bed concentration to the freeboard (cf. Figure 10). The numerical values of a obtained in the present work are plotted in the form of $a \cdot u$ as a function of the superficial gas velocity u [17] in Figure 15. The results may be summarized by

$$a \cdot u = \begin{cases} 2.4 \ \ 1/s & \text{for ash,}\ \ d_{ps} = 120\ \mu m \\ 4.0 \ \ 1/s & \text{for sand,}\ \ d_{ps} = 240\ \mu m \end{cases} \qquad (15)$$

The concentration profiles in the bed and freeboard regions are connected by the condition

$$c_{v,bed} = c_{v,freeboard} \qquad \text{for } h = h_B\,.$$

In practical cases, the total mass of solids in the reactor vessel, m_s, may be known from e.g. the total pressure drop over bed and freeboard. The total riser pressure drop is not influenced by acceleration effects. Therefore it holds

$$m_s = \int_0^{h_B} \rho_s\, c_{v,bed}(h)\, A_t\, dh + \int_0^{h_B} \rho_s\, c_{v,freeboard}(h)\, A_t\, dh \qquad (16)$$

Equation (16) may be used for the determination of the height h_B of the expanded bed.

Measured and calculated axial solids concentration profiles for sand and CFBC ash are compared in Figure 16 and 17. The agreement between experiments and calculation is quite good. The calculation scheme developed here is presently used in the authors' group for the prediction of the solids concentration distribution in the bottom zone of CFB combustors and catalytic reactors [18].

Measurements and interpretation of velocities and dimensions of void elements

For the calculation of solids volume concentrations in the bed region it has been assumed that the void phase consists of bubbles. This assumption is now examined in detail.

Local average void velocities u_v were obtained by cross-correlation of the signals obtained from two optical probes located vertically above each other. As has been described in the preceding section the local time-averaged void volume fraction was obtained as the ratio of time during which $c_v < 0.15$ was observed to the total measuring time. This signal-cutting procedure has also been used to measure local average durations t_v of void periods. A multiplication of t_v with the corresponding average void velocity u_v yielded the local average void pierced length l_v. Measured values of u_v, t_v and l_v are listed in Table 2.

l_v is a measure for a local average void dimension. If a spherical shape is assumed and if the voids are rising randomly distributed with respect to the locally measuring probe the local average sphere diameter L_v is given by

$$L_v = 1.5\ l_v \qquad (17)$$

[19]. The values of L_v which are included in Table 2 vary between 0.18 and 0.47 m. The bed diameter is 0.6 m and the probe was located at r = 0.1 m or 0.2 m from the vessel wall during these measurements. The ratio of void size to bed size or distance from the wall respectively, is so large that in most cases it is not possible for the voids to be randomly distributed with respect to the probe. Thus L_v from Equation (17) may be considered as an upper limit for an estimation of the local average void size. The lower limit is given by the average pierced length itself. In Figure 18 these upper and lower limits of the local average void size are compared with local average bubble sizes calculated from Equation (11) using eqns. (7) - (10) and (12). The calculated bubble sizes are seen to fall between the limits of the experimental void sizes.

In Figure 19 the measured local mean velocities u_v of the rising voids are compared with local average bubble rise velocities u_b calculated from Equation (9) with the bubble diameter d_v taken from Equation (11). Although there is some scatter a fair agreement between u_v and u_b is observed.

The fact that a calculation procedure which has been developed for the prediction of bubble growth in bubbling fluidized beds is able to predict void dimensions in turbulent fluidized beds with reasonable accuracy together with the finding that the voids in turbulent fluidized beds are rising at velocities which may again be described by bubble rise velocities may now finally be taken as a justification for the assumption in section 3.3 that the void phase in the turbulent fluidized bed consists of bubbles.

Recently, Bi and Fan [20] on the basis of an extensive literature survey concluded that the turbulent regime in effect exists only for Archimedes numbers Ar < 125. Beds with Ar > 125 are characterized by the existence of large gas bubbles or slugs at higher gas velocities. Since the solid materials used here have Ar numbers of 166 and 1328 (cf. Table 1), respectively, it was examined whether an existing correlation for the rise velocity of slugs could also describe the measured void velocites. For a slugging fluidized bed with continuous introduction of gas the slug rise velocity u_{sl} is given by [21, 22]

$$u_{sl} = u - u_{mf} + 0.35 \sqrt{g d_t} \qquad (18)$$

The comparison between the measured void velocities u_v and the corresponding slug rise velocities u_{sl} calculated from equ. (18) is given in Figure 20 which clearly shows that the measured void velocities cannot be interpreted as the velocities of rising slugs.

CONCLUSIONS

The following conclusions may be drawn from the present investigation:

– The solids volume concentration obtained from pressure drop measurements, $c_{v,\Delta p}$, tends to overestimate the solids holdup in the bottom region of the fluidized bed.

- Contrary to other authors' findings who mostly measured pressure fluctuations in beds of smaller diameters the present investigation in two pilot-scale fluidized beds has shown no significant changes in the fluidization characteristics when the gas velocity was increased from the bubbling to the turbulent fluidization regime. Basically the same two-phase structure with gas voids in a dense suspension was always observed. At all fluidization conditions the voids were rising at velocities which could be interpreted as the velocities of rising bubbles. These velocities could not be interpreted by the well-known Stewart-Davidson relationship for slug rise velocity. An existing bubble coalescence model was found to be sufficient to describe the sizes of the voids.

- A calculation procedure is suggested for the prediction of the axial solids volume concentration profile which includes some newly developed correlations for the dense phase voidage and for the visible bubble flow. The calculated profiles of the solids concentration are shown to provide a good description of the measurements.

NOTATION

a	decay constant, def. by eq. (14), m^{-1}
Ar	Archimedes number, def. by eq. (5), -
A_t	bed's cross-sectional area, m^2
c_v	solids volume concentration, -
$c_{v\infty}$	limiting solids volume concentration for $h \rightarrow \infty$, cf. eq. (14), -
d_{ps}	Sauter diameter, m
d_t	bed diameter, m
d_v	bubble volume equivalent diameter, m
d_{vo}	initial bubble diameter from eq. (12), m
D	relative dielectric constant, -
g	acceleration due to gravity, ms^{-2}
h	height above distributor plate, m
h_B	expanded bed height, m
H_{mf}	bed height at minimum fluidization, m
l_v	local mean void pierced length, m
L_v	local mean void size, def. by eq. (17), m
m_s	bed mass, kg
$p(c_v)$	probability density of local solids volume concentration, -
Q_3	cumulative mass distribution, -
r	radial distance from vessel centerline, m
t_v	local average duration of void period, s
Re_p	Reynolds number, def. by eq. (4), -
u	superficial fluidizing velocity, m/s
u_b	local average bubble rise velocity, m/s
u_{mf}	minimum fluidizing velocity, m/s
u_{sl}	slug rise velocity , m/s
u_t	terminal velocity of single particle, m/s
u_v	local average velocity of voids, m/s
U	voltage, V
U_o	voltage corresponding to zero local solids volume concentration, V
$\dot{V}_b$	visible bubble flow, $m^3\, m^{-2}\, s^{-1}$
$\dot{V}_o$	gas flow issuing through an individual opening in the gas distributer, cf. eq. (12), $m^3 s^{-1}$
ε	voidage, -
ε_b	bubble volume fraction, -
φ	parameter, defined by eq. (8), -
ν	kinematic viscosity, $m^2\, s^{-1}$
ϑ	parameter, defined by eq. (10)
ρ_s	solids density, $kg\, m^{-3}$

indices:

d	dense phase
f	fluid
fb	fixed bed
m	mixture
mf	minimum fluidization
s	solid

LITERATUR CITED

[1] Brereton, C.M.H. and Grace, J.R., TransIChemE Vol. 70, Part A, 1992, 246/251.

[2] Andersson, S., Johnsson, F. and Leckner, B., Proc. 10th Int. Conf. Fluidized Bed Combustion, ASME, New York 1989, 239/247.

[3] Johnsson, F., Svensson, A. and Leckner, B. in: Fluidization VII (Potter, O.E. and O.J. Nicklin (Eds.)), Engineering Foundation, New York 1992, 471/478.

[4] Svensson, A., Johnsson, F. and Leckner, B., Proc. 12th Int. Conf. Fluidized Bed Combustion, ASME, New York 1993, 887/897.

[5] Werther, J. and Molerus, O., Int. J. Multiphase Flow 1(1973), 103/122.

[6] Bakker, P.J. and Heertjes, P.M., British Chem. Eng. 4(1959), 534/539.

[7] Wiener, O., Abhandl. Sächs. Ges. Wiss. Math. Phys,. Kl. 32(1912), 509/598.

[8] Rensner, D. and Werther, J., Part. Part. Syst. Charact. 10(1993), 48/55.

[9] Werther, J. and Molerus, O., Int. J. Multiphase Flow 1(1973), 123/138.

[10] Clift, R. in Geldart, D. (Ed.): Gas Fluidization. John Wiley, Chichester 1986, p.53.

[11] Hilligardt, K. and Werther, J., Proc. World Congress III of Chem. Engng. 1986 Tokyo/Japan Vol. III, 429/432.

[12] Khattab, I.A., Kuroda, C. and Ishida, M., J. Chem. Engng. Japan 21(1988), 282/287.

[13] Werther, J. in Ullmann's Encyclopedia of Industrial Chemistry Vol. B4, VCH Publishers, New York 1992, p. 239.

[14] Werther, J., Ger. Chem. Eng. 1(1978), 166/174.

[15] Hilligardt, K. and Werther J., Ger. Chem. Eng. 9(1986), 215/221.

[16] Lewis, W.K., Gilliland, E.R. and Lang, P.M., Chem. Eng. Progr. Symp. Ser. 58(38), 1962, 65/77.

[17] Kunii, D. and Levenspiel, O., Fluidization Engineering, Butterworth-Heinemann, Boston, 1991.

[18] Schoenfelder, H.., Werther, J., Hinderer, J. and Keil, F., 1993 AIChE Annual Meeting, St. Louis, Mo.

[19] Werther, J., Trans. Instn. Chem. Engrs. 52(1974), 149/159.

[20] Bi, H. and Fan, L.S., AIChE J. 38 (1992) 2, 297/301.

[21] Stewart, P.S.B. and Davidson, J.F., Powder Technol. 1 (1967), 72.

[22] Ormiston, R.M., Mitchell, F.R.G. and Davidson, J.F., Trans.Inst. Chem. Engrs. 43 (1965), T 209/220.

Table 1 Solids properties

	quartz sand	CFBC bed ash
density ρ_s, kg/m^3	2600	2800
Sauter diameter d_{ps}, µm	240	120
Archimedes number Ar,-	1328	166
minimum fluidization velocity u_{mf}, m/s	0.13	0.033
void fraction at minimum fluidization ε_{mf}, -	0.45	0.45
relative dielectric constant D_s, -	7.0	14.7
terminal velocity for Sauter diameter $u_t(d_{ps})$, m/s	1.92	0.85
range of fluidization velocities u, m/s	0.5 - 3.0	0.4 - 2.1

Table 2 Local average void velocities and dimensions (bed diameter 0.6 m, probe located at r = 0.1 m, h = 0.38 m (sand) and 0.34 m (ash)

bed material	sand			ash		
u, m/s	0.5	2.1	3.0	0.4	1.0	1.4
u_V, m/s	1.5	2.1	2.4	1.5	2.5	3.1
t_V, s	0.11	0.12	0.13	0.078	0.090	0.083
l_V, m	0.17	0.26	0.31	0.12	0.225	0.26
L_V, m	0.25	0.39	0.47	0.18	0.34	0.39

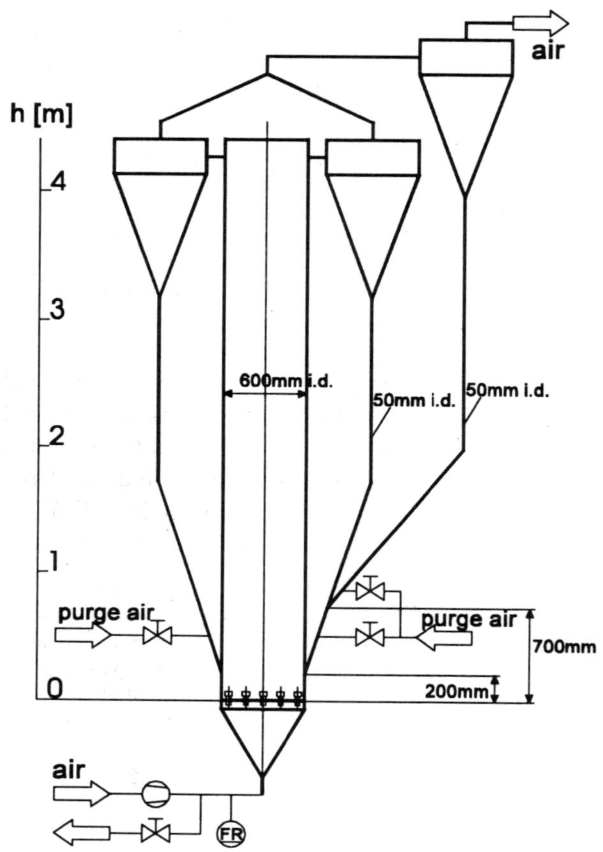

Figure 1 600 mm diameter fluidized bed unit.

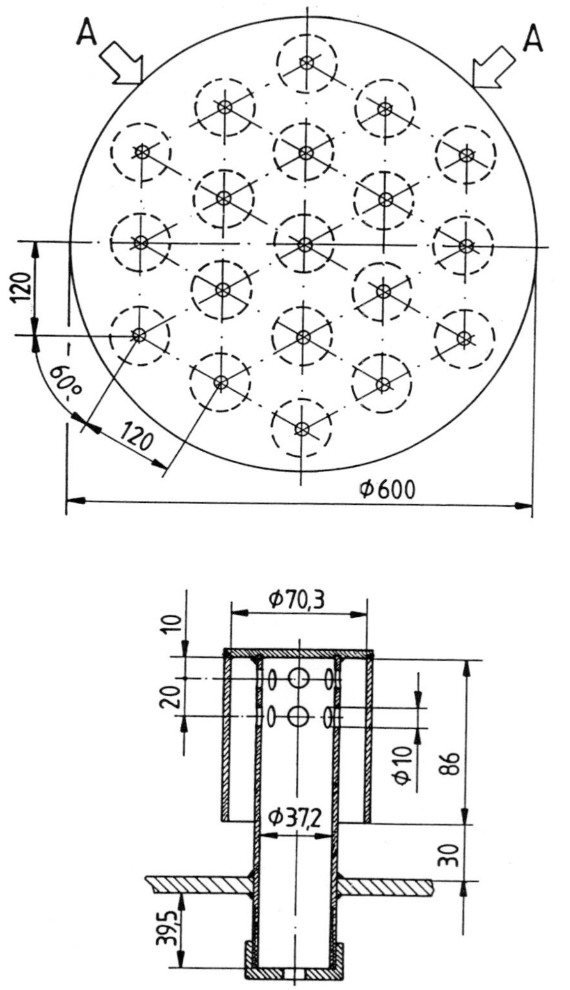

Figure 2 Gas distributor design (A = position of solids return leg, dimensions in mm).

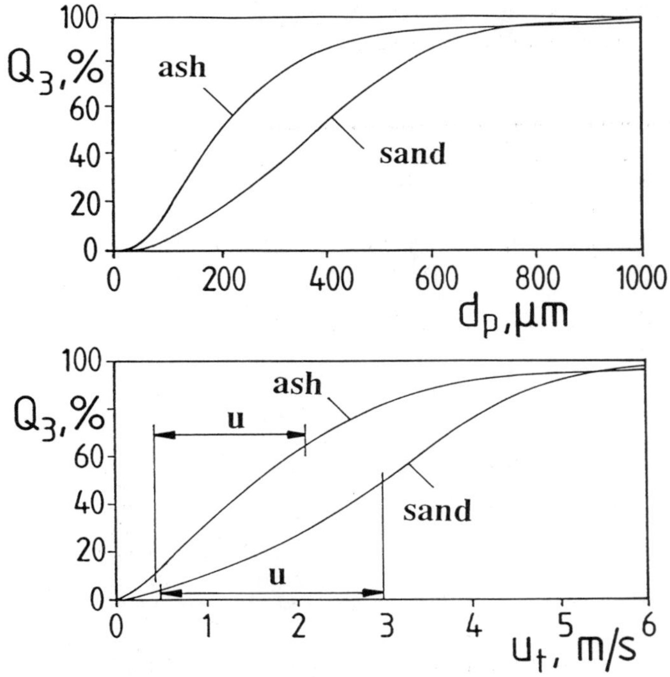

Figure 3 Bed particle size distribution and cumulative mass distribution of bed particles' terminal velocities with indication of range of fluidizing velocities u.

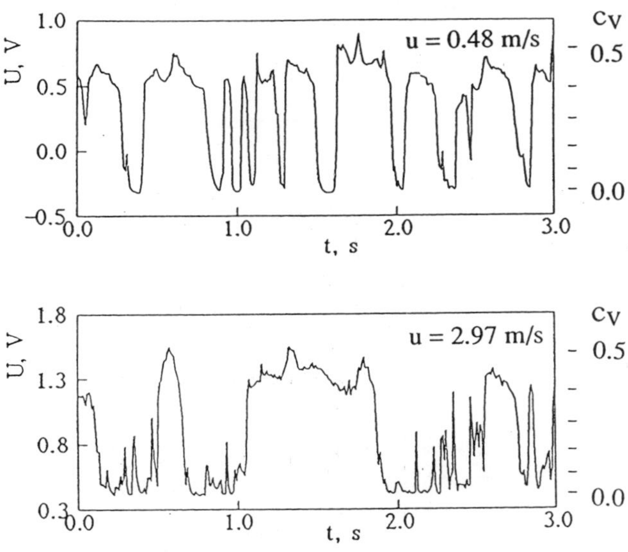

Figure 7 Capacitance probe signals (0.6 m dia. bed, sand, H_{mf} = 0.64 m, h = 0.34 m, r = 0.1 m).

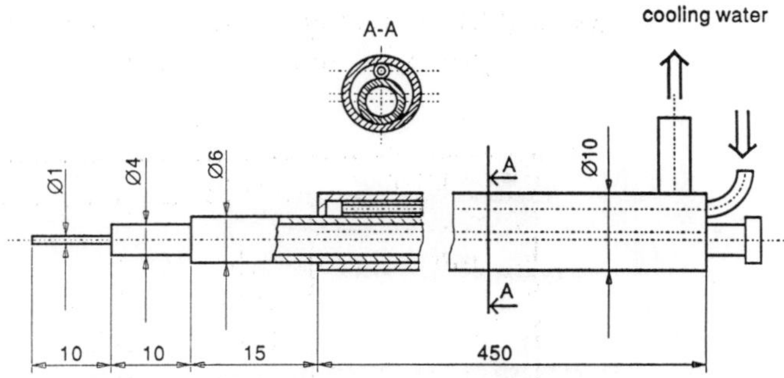

Figure 4 Water-cooled capacitance probe (dimensions in mm).

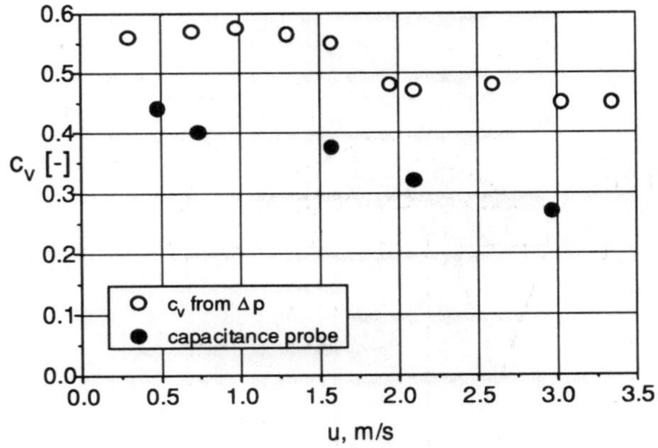

Figure 5 Comparison of cross-sectional average solids volume concentrations c_v obtained from pressure drop (between heights 0.34 m and 0.44 m above distributor) and those obtained from capacitance probe measurements at h = 0.34 m (0.5 x 0.5 m^2 square bed, sand, H_{mf} = 0.64 m).

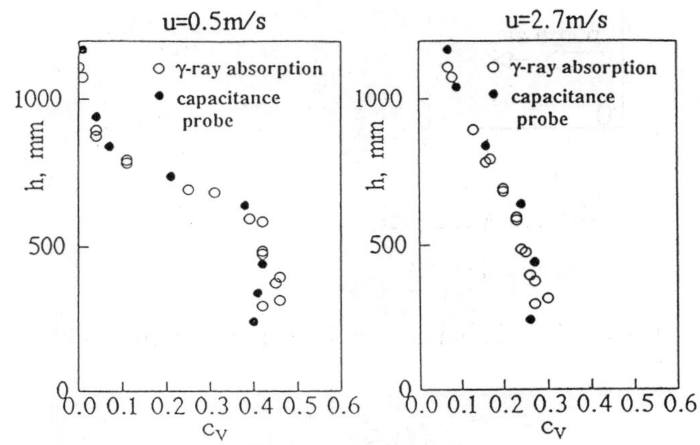

Figure 6 Comparison of solids concentrations from γ-ray measurements with those obtained by averaging local capacitance probe measurements along a traverse through the bed (0.5 x 0.5 m^2 square bed, sand, H_{mf} = 0.62 m).

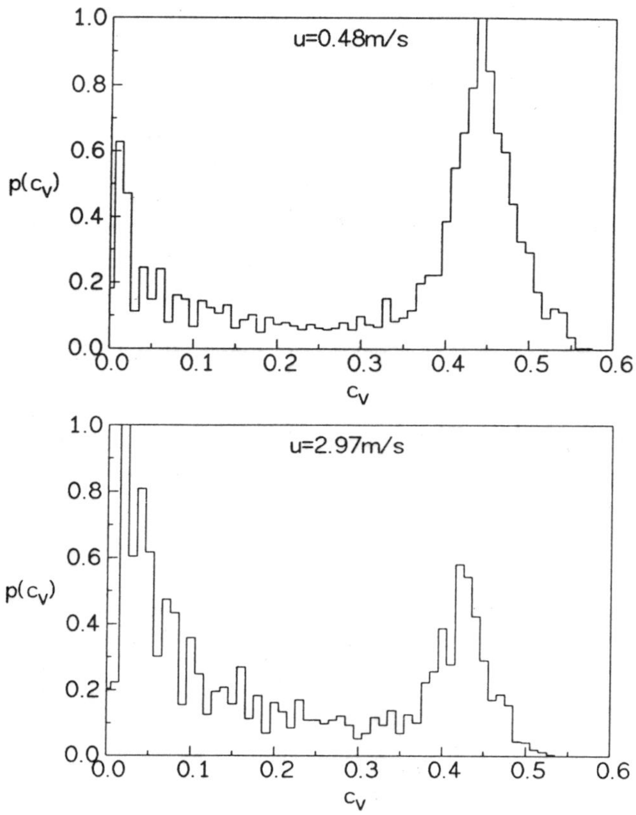

Figure 8 Probability density distributions of the signals shown in Figure 7.

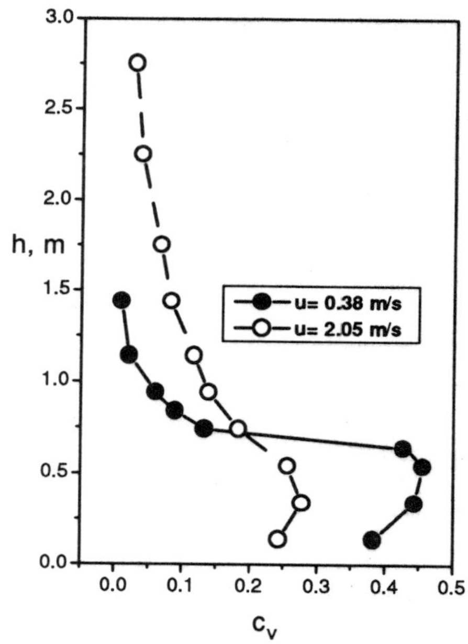

Figure 10 Axial profiles of cross-sectional average solids volume concentration (0.6 m dia. bed, ash, $H_{mf} = 0.65$ m).

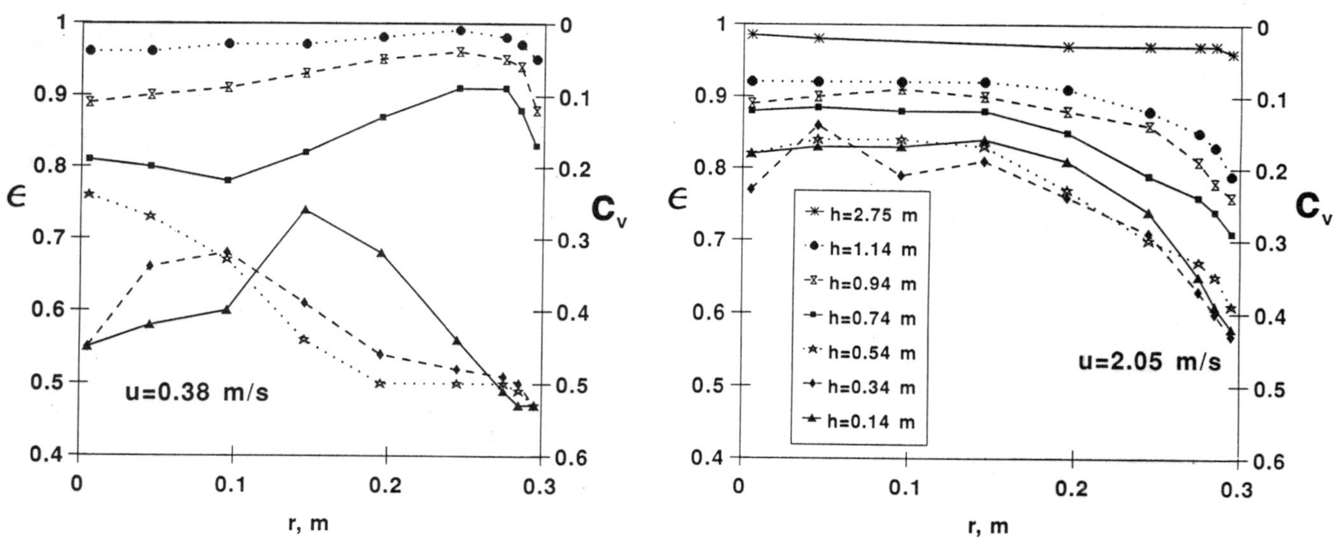

Figure 9 Horizontal distributions of local voidage ε (0,6 m dia. bed, ash, $H_{mf} = 0.65$ m).

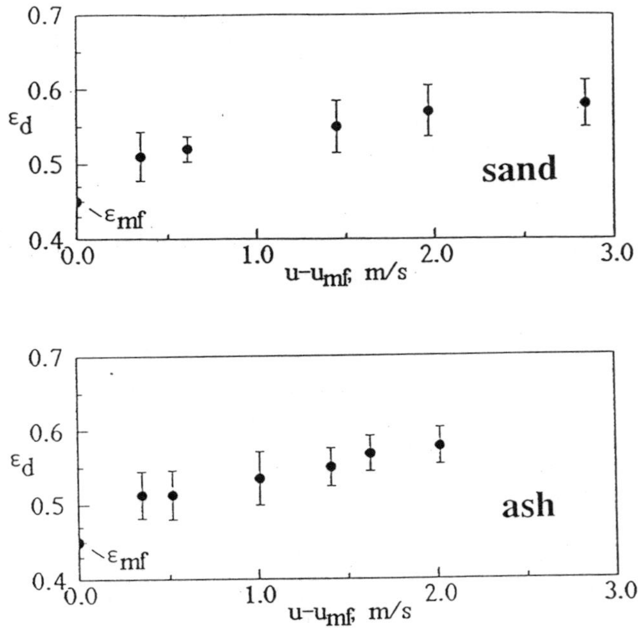

Figure 11 Dense phase voidage ε_d as a function of the excess gas velocity ($u - u_{mf}$) 0.6 m diameter bed).

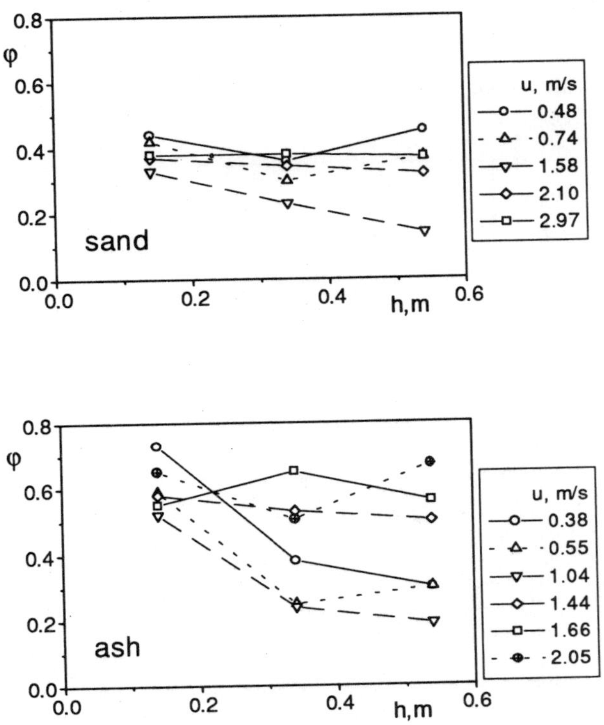

Figure 13 The parameter φ as a function of height, gas velocity and particle properties 0.6 m diameter bed).

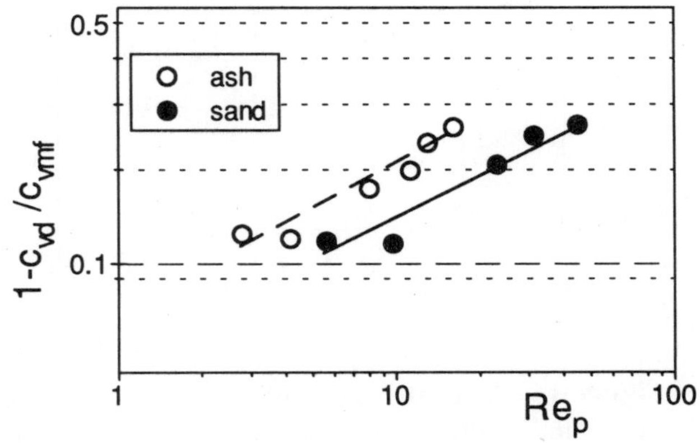

Figure 12 A comparison of the model equation (6) with measured values of the dense phase concentration c_{vd}.

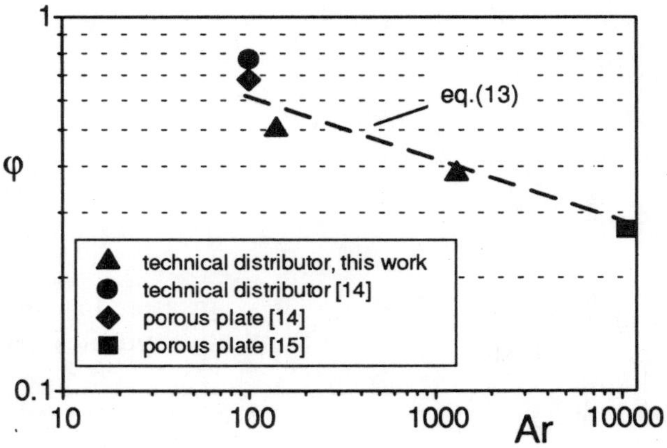

Figure 14 Dependence of φ on Archimedes number.

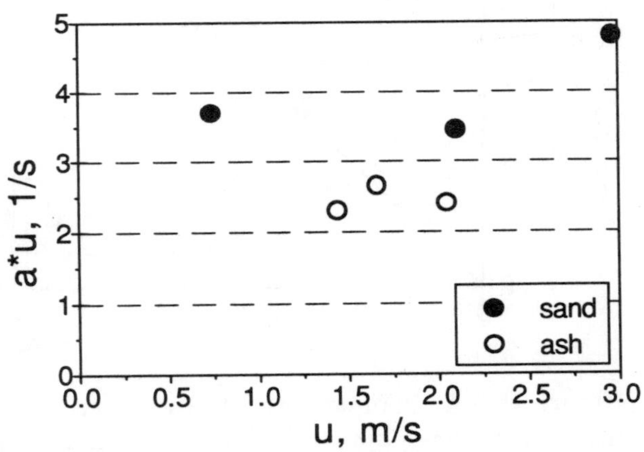

Figure 15 Influence of superficial gas velocity on the product $a \cdot u$.

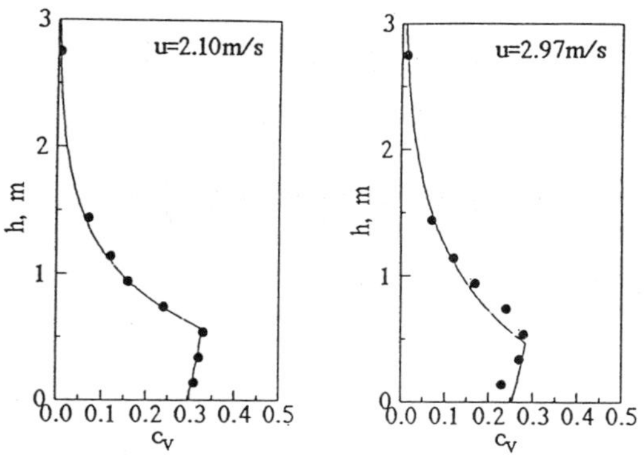

Figure 16 Comparison of measured and calculated axial solids concentration profiles for sand.

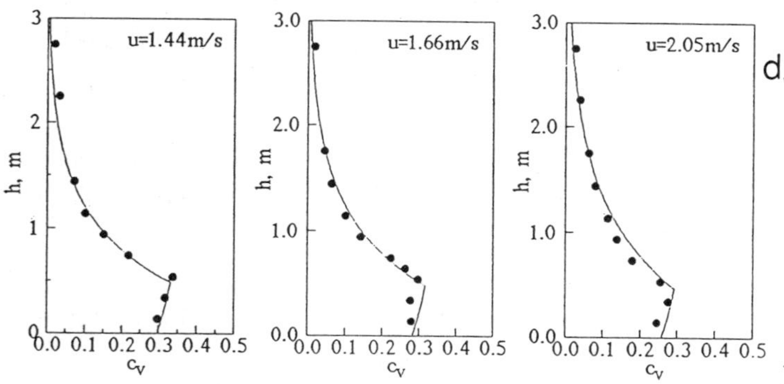

Figure 17 Comparison of measured and calculated axial solids concentration profiles for ash.

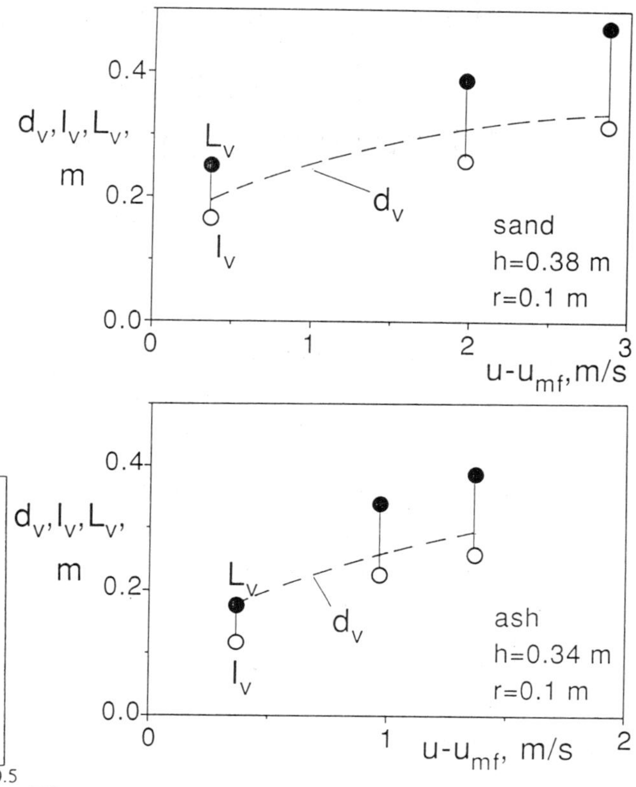

Figure 18 A comparison of upper (L_v) and lower (l_v) limits of local average void sized with calculated bubble diameters d_v (bed diameter 0.6 m).

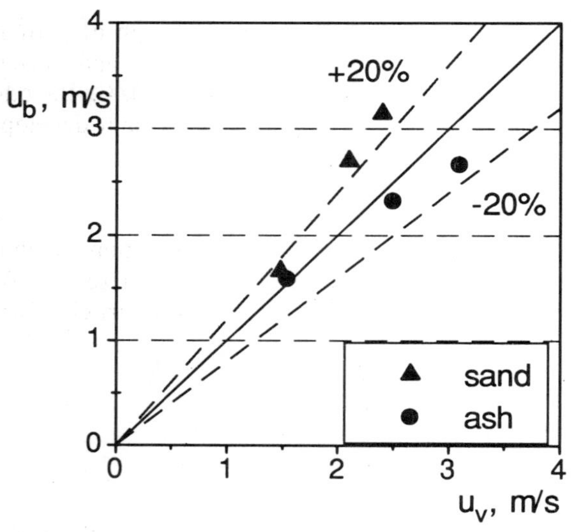

Figure 19 A comparison of measured local mean void velocities u_v with calculated local average bubble rise velocities u_b (bed diameter 0.6 m).

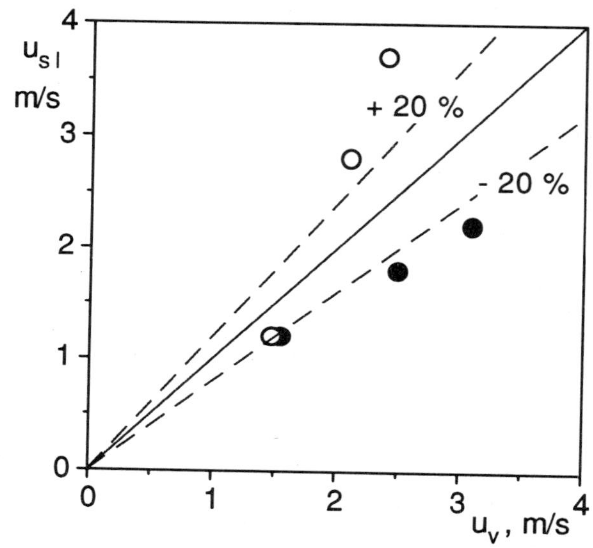

Figure 20 A comparison of measured local mean void velocities u_v with rise velocities of slugs calculated from Equation (18) (bed diameter 0.6m).

The Introduction and Measurement of Rheological Parameters for the Characterization of Emulsion Phase in Fluidized Beds

H.O. Kono, Y. Itani, E. Aksoy

Chemical Engineering Department, West Virginia University, P.O.Box 6101, Morgantown, WV 26506

The emulsion phase properties of fine powders play an important role in determining the characteristics of fluidization. In this study, two important rheological parameters, the plastic deformation coefficient (Y) and the fracture strength (σ_f) of uniformly aerated powders, which can represent the property of emulsion phase, were defined and experimentally measured within the gas velocity range of $U_{mf} \leq U \leq U_{mb}$ where during this period, the powder structure was assumed to be perfectly homogeneous.

Four different sample powders, such as, spent FCC, glass beads, starch, and carbon black were used. The results show that there exists a clear correlation between the bubble characteristics, for example, size and shape and the proposed rheological parameters.

INTRODUCTION

Since the fluidization property of fine powders (d_p = 10 ∼ 100 μm) plays a significant role in determining the bubble behavior in fluidized beds, the characterization of the emulsion phase has been investigated by several investigators.

Historically, Rietema (1) has first indicated the importance of the structure of fluidized particles in terms of interparticle forces. Later, Mutsers et al. (2) has defined the concept of the rigidity of particles in fluidized beds, even though their approach was completely different from what is going to be discussed here.

To characterize the emulsion phase, some researchers (3,4) applied the concept of the tensile strength of packing particles under a specified packed bed condition. The disadvantage of the above method is that these tensile strengths were measured under a packed bed condition with the voidage much lower than that of at the minimum bubbling point, and that the properties of the actual emulsion phase were estimated from the extrapolation from packed bed data, taking the voidage change into consideration.

Another approach came from Kono (3), in which the plastic deformation coefficients of some sample powders were calculated. However, the tensile strengths in that work were also estimated through the extrapolation.

The concept of the tensile strength of particles was first introduced by Rumpf (5), and several measurement methods were developed (6). On the other hand, the shear stress test was defined and developed by Jenike (7).

These conventional characterization methods were found to be effective (1) to characterize very fine or ultra fine powders with high interparticle forces under packed bed conditions and, (2) to get the direct shear test on relatively coarse particles under a high loading shear condition such as in big silo hoppers. The disadvantages, however, are (1) the split-cell tensile strength method could not be utilized directly to characterize the powder properties for the fluidization, since that method could not be applied for aerated powders, and, (2) the accuracy of the split-cell method is not satisfactory. For this reason, the accuracy of the methods proposed by Rumpf (8), Jimbo (6), Kono (9) and Khoe (4) are not always satisfactory enough for the prediction of fine powder fluidization.

In this study, two important rheological parameters, namely, the plastic deformation coefficient (Y) and the fracture strength (σ_f) of uniformly aerated powders, which can represent the property of the emulsion phase, were defined and experimentally obtained within the gas velocity range of $U_{mf} \leq U \leq U_{mb}$. The properties of the aerated powder defined above were used as criteria for the characterization. The tensile strength of the aerated powder at $U = U_{mb}$ was defined as $\sigma_{f,mb}$ at which the first fracture occurs in the homogeneous powder structure. The correlation between these rheological parameters and the fluidization characteristics of fine powders showed pretty good agreement.

BASIC CONSIDERATION

The bubble pictures of four different materials in the transparent two-dimensional fluidized beds together with the fracture strengths ($\sigma_{f,mb}$) and plastic deformation coefficients (Y) of the aerated powder are shown in Figure 1. Experimental conditions of these pictures are: starch, $U_o/U_{mf}=15$; glass beads, $U_o/U_{mf}=13$; carbon black, $U_o/U_{mf}=16.2$; spent FCC, $U_o/U_{mf}=14.8$. As is seen, there exists considerable differences in shapes and sizes of these bubbles for each different powders with different $\sigma_{f,mb}$ and Y.

The following physical changes should occur in the bed of sample powders in the range of the defined aerated condition of $U_{mf} \leq U \leq U_{mb}$.

The reasonable assumption made in the following discussions is that the aerated powders in this range of the gas velocity should form a completely homogeneous structure. This structure will expand between U_{mf} and U_{mb} while still maintaining the homogeneous packing structure. At U_{mb}, however, this structure will start being fractured generating the first initiation of the bubble.

When the gas velocity reaches the minimum fluidization velocity (U_{mf}), the gravitational force should be compensated by the drag force of the flowing gas. The pressure drop across the packed bed of fine powder at this minimum fluidization condition can approximately be expressed by the viscous term of Ergun's equation (see Figure 2),

$$\Delta P = 150\, H \frac{(1-\varepsilon)^2}{\varepsilon^3} \frac{\mu U}{(\phi_s d_p)^2} \tag{1}$$

When we increase further the gas velocity U from U_{mf} to U_{mb}, the bed height will expand H_{mf} to H_{mb}, during which the bed voidage will change between ϵ_{mf} and ϵ_{mb} (See Figure 2b). The pressure drop across the bed actually remains constant in this range.In order to introduce the concept of the principle of virtual work, let us make the following assumptions and theoretical considerations: (See Figure 2)

1) Under the actual experimental condition the voidage changes from ϵ_{mf} to ϵ_{mb}. Therefore the pressure drop across the bed remains constant at $U_{mf} \leq U \leq U_{mb}$.

$$\Delta P_D = 150\, H_{mf} \frac{(1-\varepsilon_{mf})^2}{\varepsilon_{mf}^3} \frac{\mu U_{mf}}{(\phi_s d_p)^2}$$

$$= 150\, H_{mb} \frac{(1-\varepsilon_{mb})^2}{\varepsilon_{mb}^3} \frac{\mu U_{mb}}{(\phi_s d_p)^2} \tag{2}$$

2) If the bed voidage remains the same as ϵ_{mf} during the gas velocity U increase from U_{mf} to U_{mb}, the pressure drop across the bed (ΔP_H) should be as follows:

$$\Delta P_H = 150\, H_{mf} \frac{(1-\varepsilon_{mf})^2}{\varepsilon_{mf}^3} \frac{\mu U_{mb}}{(\phi_s d_p)^2} \tag{3}$$

3) Maintaining the same assumption of 2, the difference of (ΔP_H - ΔP_D) was defined as $\Delta P^*(U)$ as shown in Figure 2a:

$$\Delta P^*(U_{mb}) = \Delta P_H(U_{mb}) - \Delta P_D = 150\, H_{mf}$$

$$\frac{(1-\varepsilon_{mf})^2}{\varepsilon_{mf}^3} \frac{\mu}{(\phi_s d_p)^2} (U_{mb}-U_{mf}) \tag{4}$$

and when $U=U_{mf}$, $\Delta P^*(U_{mf})=0$. Note that, this excess pressure drop is a function of gas velocity.

Now, our following discussion will be made per unit area of powder layer.

When the gas velocity is increased from U_{mf} to U_{mb}, the excessive gas flow rate (U_{mb} - U_{mf}) passes through the expanded bed with the pressure drop of ΔP_D (See Figure 2a), increasing the bed voidage from ϵ_{mf} to ϵ_{mb}, and causing the bed expansion from H_{mf} to H_{mb}. The excessive energy (E_D), caused by the excessive gas flow of (U_{mb} - U_{mf}) per unit layer of powder in the bed should be [(1/2) ($\Delta P^*(U_{mb})/N$) (U_{mb} - U_{mf})].

4) The unit layer consists of two adjacent layers contacting as shown in Figure 3c. Even though the structure of the unit layers shown here is very schematic, the voidage of the unit layer can really be changed in a pretty wide range (10).

The excessive energy (E_D) per unit time for unit powder layer should be as follows: (See Figure 3a)

$$E_D = \frac{1}{2}\frac{\Delta P^*(U_{mb})}{N}(U_{mb}-U_{mf})$$

$$= \frac{\Delta P^{**}(U_{mb})}{2}(U_{mb}-U_{mf}) \qquad (5)$$

where $N = H_{mf} / d_p$ and $\Delta P^{**} = \Delta P^*(U_{mb}) / N$. Here, N is by definition the number of unit layers of the powder structure at the minimum fluidization point.

5) As it is known, the bed voidage was actually changed from ϵ_{mf} to ϵ_{mb} and the bed height was also changed from H_{mf} to H_{mb}. The excessive energy to cause the expansion of the powder structure per unit layer can be expressed by Equation (8) (See Figure 3b):

$$E_\sigma = \frac{1}{2}(\sigma_{f,mb}+\sigma_{f,mf})(U_{mb}-U_{mf}) \qquad (6)$$

where ($\sigma_{f,mf}$) and ($\sigma_{f,mb}$) are the fracture strengths of fine powder packing structure at U_{mf} and U_{mb} respectively.

The general tensile strength equation of the powder structure in a packed bed (σ_t) was defined by Rumpf (5) as follows:

$$\sigma_t = \frac{1-\varepsilon}{\pi}k\frac{F}{d_p^2} \doteq \frac{(1-\varepsilon)}{\varepsilon}\frac{F}{d_p^2} \qquad (7)$$

where, σ_t is the tensile strength of powder structure defined per unit area measured under the packed bed condition. Since this is a qualitative equation for the sake of understanding the phanomena, the accuracy of Equation (7) is not always reliable because of the unmeasurable parameters. Accordingly, we did not use Equation (7), but measured the fracture strength experimentally using the minimum bubbling point as the criteria of the initiation of powder structure fracture, as explained earlier.

6) Based upon the considerations of 1 - 5, using the principle of virtual work, E_D, *i.e.* Equation (5), should be equal to E_σ, *i.e.*, Equation (6). Therefore,

$$E_D = E_\sigma \qquad (8)$$

E_D and E_σ can also be written as follows;

$$E_D = \int_{U_{mf}}^{U_{mb}} \Delta P^{**}(U)\ dU = \frac{1}{2}\Delta P^{**}(U_{mb})(U_{mb}-U_{mf}) \qquad (9)$$

$$E_\sigma = \int_{U_{mf}}^{U_{mb}} \sigma_f(U)\ dU = \frac{1}{2}(\sigma_{f,mb}+\sigma_{f,mf})(U_{mb}-U_{mf}) \qquad (10)$$

By combining Equations (9) and (10) together with Equation (8) and (7), the fracture strength at the minimum bubbling point ($\sigma_{f,mb}$) can be derived as,

$$\sigma_{f,mb} = \frac{\Delta P^{**}}{\left[1+\left(\frac{\varepsilon_{mb}}{\varepsilon_{mf}}\right)\left(\frac{1-\varepsilon_{mf}}{1-\varepsilon_{mb}}\right)\right]} \qquad (11)$$

Here, the assumption was made that the change of $\sigma_{f,mb}$ with respect to U should be approximately linear.

The second rheological parameter that will be introduced here is the plastic deformation coefficient (Y) of the powder structure. Figure 4 shows the deformation of the powder bed at U_{mf} and U_{mb}. Based upon the discussions given previously, the strain of the Unit powder layer can be written as:

$$S = \frac{N(h_{mb}-h_{mf})}{N\,h_{mf}} = \frac{H_{mb}-H_{mf}}{H_{mf}} \qquad (12)$$

As previously discussed, we assumed the homogeneous powder structure. Then, the strain in a unit powder layer is the same as that of the whole emulsion phase. Therefore, the plastic deformation coefficient of the aerated powder structure for a single unit powder layer was defined in Equation (13),

$$Y = \frac{\dfrac{\Delta P^{*}}{N}}{\left(\dfrac{h_{mb}-h_{mf}}{h_{mf}}\right)} = \frac{\Delta P^{**}}{\left(\dfrac{H_{mb}-H_{mf}}{H_{mf}}\right)} \qquad (13)$$

Note that, this definition can also be extended to the whole emulsion height as already shown in Equation (12).

It is very important to note at this point that, both the fracture strength and the plastic deformation coefficient results were obtained by using experimental data.

EXPERIMENTAL

The experimental parameters, *i.e.*, ΔP^{**}, ϵ_{mb}, ϵ_{mf}, H_{mb} and H_{mf}, which are necessary data to obtain the fracture strength ($\sigma_{f,mb}$) and plastic deformation coefficient (Y) were obtained through the well-known collapse test (1,11).

The experimental apparatus is schematically shown in Figure 5. The cylindrical fluidized bed was 4" in diameter. Through the preliminary experiments, the 4" diameter fluidized bed was found to be appropriate to avoid the wall effect for collapse tests done at least at ambient temperature, using our sample powders. The fluidizing air was supplied from a compressed air line via flowmeter. A video recorder was used to record experiments. For pressure measurements, a Validyne Model P305D pressure sensor was utilized together with a personal computer for data processing. The minimum bubbling points were measured in the same way as reported in literature (1,11,12).

The observation of recordings gives bed height versus time plots. The emulsion bed heights for each flow rate is then plotted as a function of velocity. The pressure drop versus velocity curve on the other hand, is plotted through the well-known procedure. As a result, all the parameters are obtained from the experimental data. These data were utilized to calculate the rheological system parameters by using the equations given in the previous section.

As sample powders, spent FCC, glass beads, special starch, and carbon black were used. The physical properties of these are given in Table 1. Both the density and the particle size values in Table 1 were measured experimentally. A granulated form of carbon black was utilized since its primary size was 1 - 2 μ. The starch, on the other hand, is a special "flow conditioner" which has extremely good flowability. We confirmed in advance that all our sample powders could be well fluidized.

RESULTS and DISCUSSION

The experimental data of collapse tests and pressure measurements are shown in Table 2. The results of the fracture strength ($\sigma_{f,mb}$) and plastic deformation coefficient (Y) were obtained experimentally from the data of Table 2 and given in Table 3.

As mentioned earlier, the tensile strength measurement of the fluidizable powder with low σ_t employing the split-cell method used to be found practically impossible. Only for very fine powders with very high interparticle forces, the tensile strength by split-cell method were measured by Jimbo et al. (6).

Figure 6 shows the data of fracture strength of fluidizable powders at aerated condition using the proposed method. It also shows the tensile strength of very fine powders at packed bed condition, using the classical split-cell method.

As seen in Figure 6, the powder for fluidization with the pretty low fracture strength could be precisely measured by this new method. The measurable range of σ_t by the classical split-cell method was 10 - 10^4 Pa, while the measurable range by our new method was 1.0 - 10^{-2} Pa.

The establishment of this measurement method is significant because the low fracture strength powder's rheological parameters necessary for the prediction of their fluidization performance can only be measured by our proposed method. For the characterization of fluidization the fracture strength measurement at aerated condition is essential.

In Figures 7 and 8, the values of Y and $\sigma_{f,mb}$ are given for sample powders, respectively. The comparison of these values with the bubble characteristics, as shown in Figure 1, indicates that there is a very close correlation between these rheological parameters ($\sigma_{f,mb}$ and Y) and the bubble characteristics such as bubble size and shape. The pictures obtained by transparent two dimensional fluidized bed shows; when the $\sigma_{f,mb}$ values decreased from 0.130 Pa to 0.006 Pa, the bubble size also decreased significantly. The same trend can also be observed with respect to Y values, ranging from 1.276 Pa to 0.022 Pa.

CONCLUSIONS

Based upon the proposed powder structure model, the fracture strength ($\sigma_{f,mb}$) and the plastic deformation coefficient (Y) were introduced and measured to characterize the aerated emulsion phase property in the velocity range of $U_{mf} \leq U \leq U_{mb}$ in fine powder fluidized beds.

The fracture strength of powders ($\sigma_{f,mb}$) and the plastic deformation coefficient (Y) of the emulsion phase at $U_{mf} \leq U \leq U_{mb}$ were experimentally obtained using the collapse tests and other related experiments.

Finally, the pictures of bubble characteristics obtained by two dimensional transparent fluidized beds were correlated to the values of $\sigma_{f,mb}$ and Y. Within the range of these experiments, the bubble characteristics seem to be determined by these rheological parameters $\sigma_{f,mb}$ and Y.

NOTATION

d_p: Particle Diameter, [m]
E_D: Excessive Energy of Gas Flow per Unit Time per Unit Layer of Powders Defined in Equation (5), [Watt]
E_σ: Energy to Cause the Deformations of the Aerated Powder Structure per Unit Time per Unit Layer of Powders as Defined in Equation (6), [Watt]
F: Interparticle Force, [N]
H: Bed Height, [m]
k: Coordination Number of the Powder, [-]
N: Number of the Unit Layer of Powders in the Bed at U_{mf}, [-]
S: Strain Defined in Equation (12), [-]
ΔP: Pressure Drop of the Bed, [Pa]
ΔP^*: Pressure Difference Defined in Equation (4), [Pa]
U: Superficial Velocity, [m/sec]
ϵ: Voidage of the Bed, [-]
μ: Viscosity of the Gas, [Pa.sec]
ϕ: Carman's Shape Factor of the Particle, [-]
σ_f: Fracture Strength of the Powder Structure, [Pa]
σ_t: Tensile Strength of the Powder, [Pa]

Subscripts

mf: at Minimum Fluidization
mb: at Minimum Bubbling

LITERATURE CITED

1. RIETEMA, K., *Proc. Int. Symp. on Fluidization*, Ed. A. Drinkenburg, 154 (1967).

2. MUTSERS, S.M.P. and K. RIETEMA, *Powder Technology*, **18**, 239 (1977).

3. KONO, H.O., T. ELLS, S. CHIBA, M. SUZUKI and E. MORIMOTO, *Powder Technology*, **52**, 69 (1987).

4. KHOE, G.K., T.L. IP and J.R. GRACE, *Powder Technology*, **66**, 127 (1991).

5. RUMPF, H., *Chem. Ing. Tech.*, **30**, 3, 144 (1958).

6. JIMBO, G., R. YAMAZAKI, J. TSUBAKI and H. KAMIYA, "Memoirs of the Faculty of Engineering", Nagoya Univ., Japan (1988).

7. JENIKE, A.W., Ph.D. Thesis, Bull. Univ. Utah, No. 108, 52 (1961).

8. RUMPF, H., *Chem. Ing. Tech.*, **42**, 538 (1970).

9. KONO, H.O., "Transport Phenomena in Fluidized Particle Systems," Edited by L.K. Doraiswamy and A.S. Mujumdar, Elsevier Science Publishers, B.V. Amsterdam (1989).

10. MIWA, S., "Powder Technology," Nikkan Kogyo-Shinbun Publ., Tokyo (1962).

11. Abrahamsen, A.R. and D. Geldart, *Powder Technology*, **26**, 47 (1980).

12. KONO, H.O., A. S. AHMADI and M. SUZUKI, *Powder Technology*, **52**, 49 (1987).

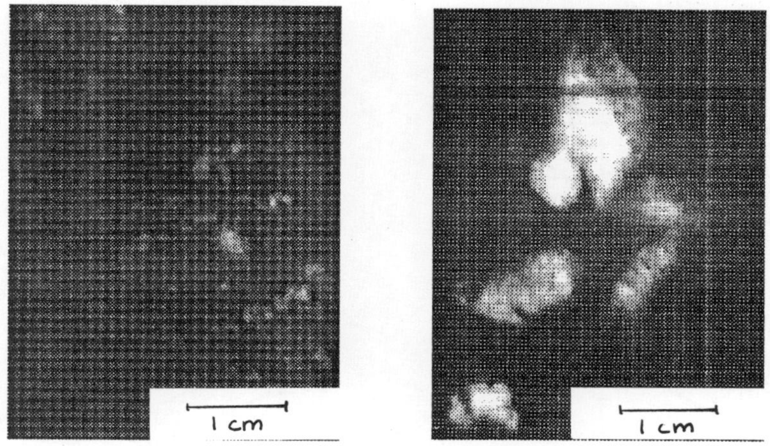

POWDER	Starch	Glass Beads
ρ_p [kg/m^3]	1,550	2,500
d_p [m]	15	30
$\sigma_{f,mb}$ [Pa]	0.006	0.026
Y [Pa]	0.022	0.285

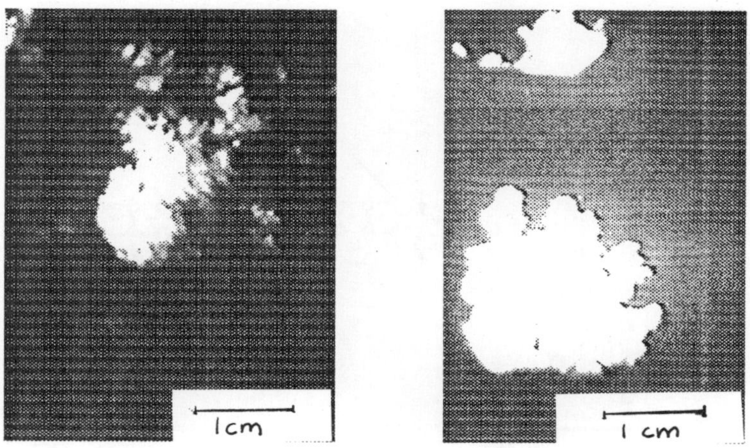

POWDER	Carbon Black	FCC
ρ_p [kg/m^3]	1,800	1,850
d_p [m]	30	65
$\sigma_{f,mb}$ [Pa]	0.077	0.130
Y [Pa]	0.311	1.276

Figure 1. Bubble pictures of sample powders in a 2-D transparent fluidized bed.

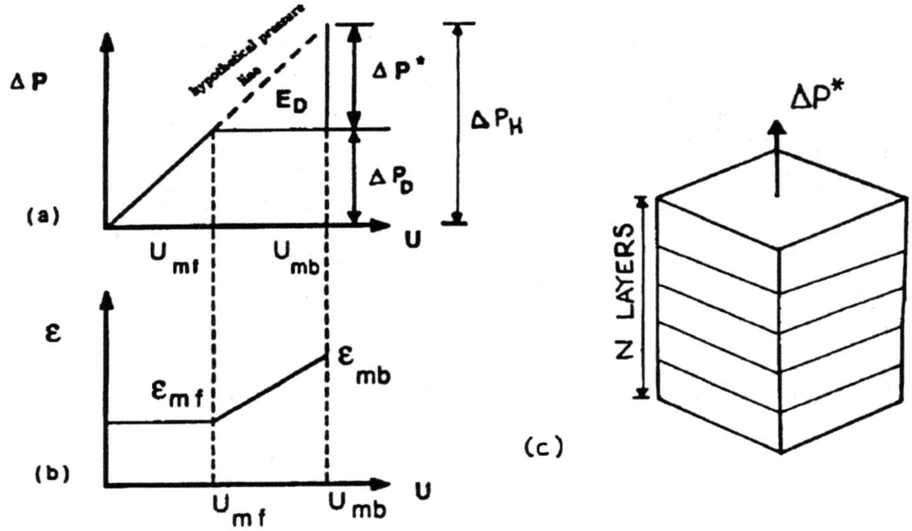

Figure 2. Physical changes occurring between U_{mf} and U_{mb} in a fine powder bed. (a) The pressure drop vs. velocity curve, (b) The bed voidage vs. velocity curve, (c) N layer representation of the powder structure.

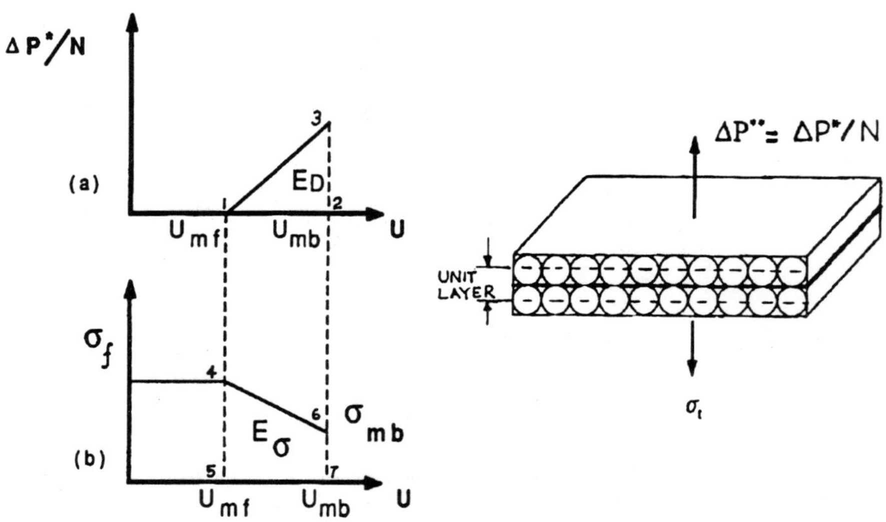

Figure 3. (a) The excessive pressure drop per layer vs. the gas velocity in the range of $U_{mf} \leq U \leq U_{mb}$. (b) The fracture strength vs. gas velocity. (c) The unit layer representation of the powder structure.

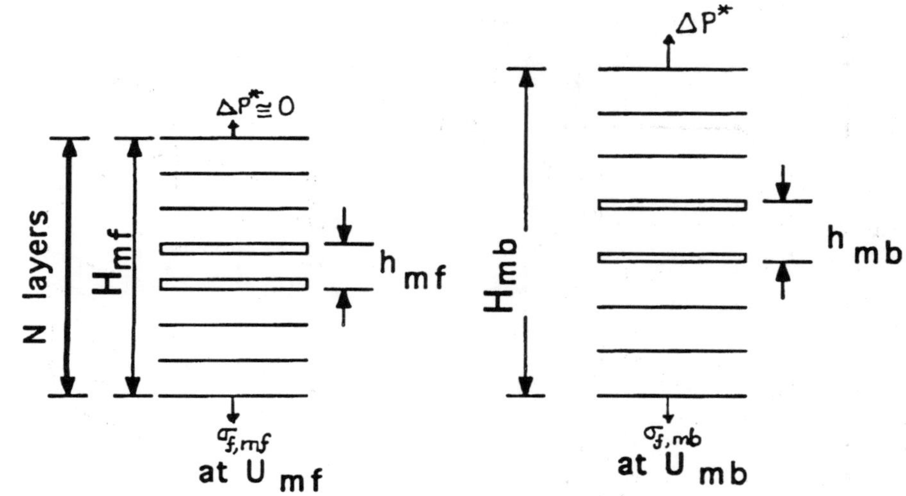

Figure 4. Schematic representation of the powder layer's expansion at U_{mf} and U_{mb} conditions.

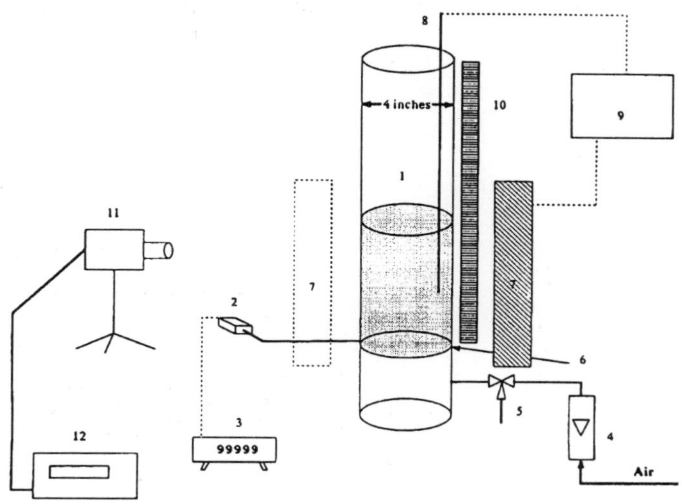

1 3-D Fluidized Bed
2 Pressure Sensor
3 Digital Multimeter
4 Flow Meter
5 3-way Valve
6 Distributor
7 Electric Heater
8 Thermo Couple
9 Temperature Control Station
10 Scale
11 Video Camcorder
12 VCR

Figure 5. Schematic diagram of experimental apparatus for the collapse test.

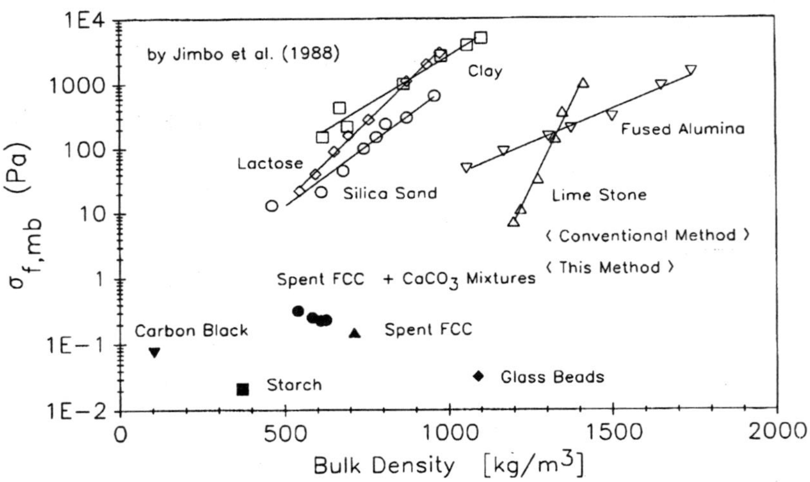

Figure 6. The comparison of $\sigma_{f,mb}$ values obtained by the new method with literature.

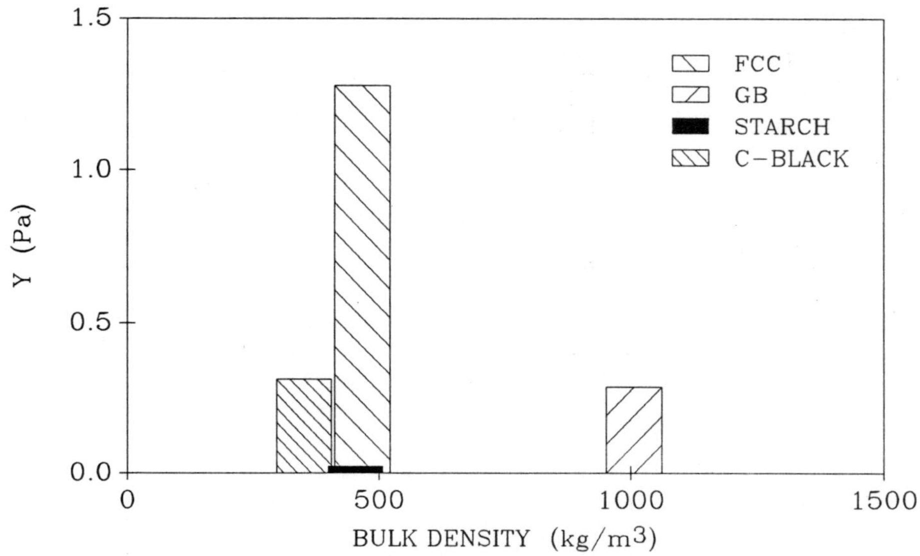

Figure 7. The plastic deformation coefficient values of sample powders as a function of bulk density.

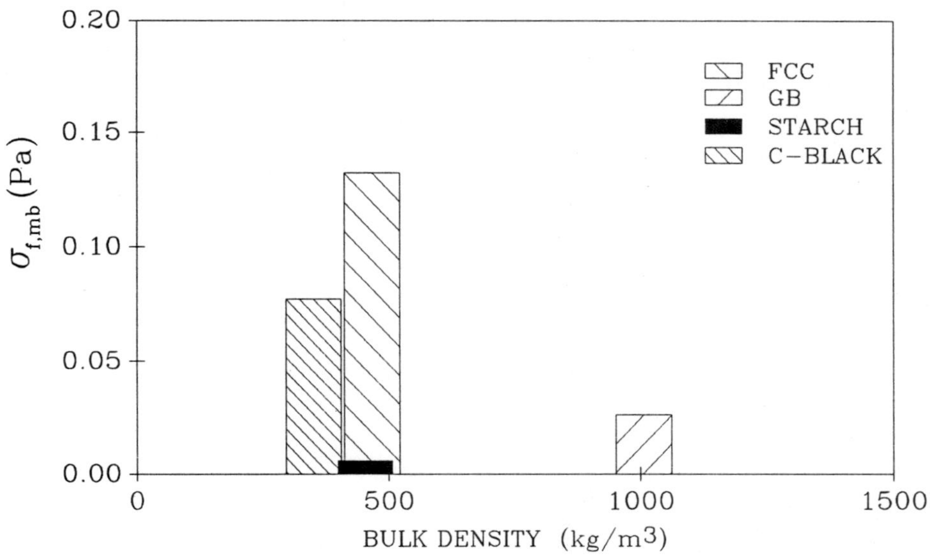

Figure 8. The fracture strength values of sample powders as a function of bulk density.

Table 1. The physical properties of sample powders at ambient temperature.

	Powder	d_p	ρ_p
		μm	kg/m^3
1	Spent FCC	65 ($20 < d_p < 125$)	1850
6	Glass Beads	30 ($10 < d_p < 60$)	2500
7	Starch	15 ($10 < d_p < 30$)	1550
8	Carbon Black	30 ($20 < d_p < 50$)	1800

Table 2. Experimental data of the collapse tests

POWDER	U_{mf}	U_{mb}	H_{mf}	H_{mb}	ρ_{mf}	ρ_{mb}	ΔP^*
	cm/s	cm/s	cm	cm	kg/m^3	kg/m^3	Pa
Spent FCC	0.61	0.90	13.8	17.3	894	713	687
Glass Beads	0.69	0.80	13.8	17.0	1341	1089	304
Starch	0.33	0.40	11.7	23.3	738	371	172
Carbon Black	0.37	1.90	14.2	23.8	174	104	995

Table 3. Experimental results of $\sigma_{f,mb}$ and Y

POWDER	$\sigma_{f,mb}$	Y
	Pa	Pa
Spent FCC	0.132	1.274
Glass Beads	0.026	0.285
Starch	0.006	0.023
Carbon Black	0.077	0.312

Bubble Trajectories and Void Fractions in the Flow Around Bluff Bodies

Andrew E. Bayly and Chris D. Rielly

Department of Chemical Engineering, University of Cambridge,
Pembroke Street, Cambridge, CB2 3RA, U.K.

A general equation of motion for a single bubble in an unsteady, non-uniform flow field is identified and contains pressure, added mass, lift and buoyancy terms. Single bubble trajectories are calculated by integrating the equation of motion over two simple potential flow solutions for flow around a cylinder. These trajectories demonstrate the important parameters affecting bubble motion, namely the slip velocity and the ratio of bubble diameter to cylinder diameter. The latter ratio is equivalent to the bubble Stokes number. Trajectories calculated using an alternative equation of motion, extrapolated from low Reynolds number theory, show little difference at lower diameter ratios, but more significant deviations for larger bubbles. Void fraction calculations reveal many similar features to published experimental results and give valuable insight to the physical effects causing bubble redistribution.

Previous research into bubbly two-phase flows has tended to concentrate on internal flows *i.e.* pipe flows, rather than external flows. In particular bubbly two-phase flow around bluff bodies has received only limited study, although it is seen to play an important role in a variety of Chemical Engineering processes, *e.g.* bubbles collecting in the wake of heat exchanger tubes in boiling two-phase cross-flows lead to poor heat transfer and "hot spots" on the tube (Lewis, 1991). A second example is found in mechanically agitated reactors and mixing vessels where the presence of a relatively small gas void fraction causes a pronounced change in the trailing vortex structure behind the impeller blades. This phenomenon leads to reduction in the power dissipated by the impeller and eventually to flooding of the blades (Van't Riet *et al.*, 1976).

The presence of bubbles is observed to have a major influence on the pressure distribution on the surface of the body, its drag coefficient and the boundary layer structure. Large spatial variations in void fraction can also occur, giving a bubble free zone near the front stagnation point on the body and high concentrations of bubbles in the wake region. Experimental studies using bubbly air/water flows around a cylinder have been reported by Inoue *et al.* (1986), Yokosawa *et al.* (1986 a & b) and Watanabe *et al.* (1990). These groups of workers found a significant change in the pressure distribution around the cylinder, and a large reduction in cylinder drag, on the introduction of less than 1 % volume fraction of air bubbles, to a flow in the sub-critical range. These changes were very similar to the sub-critical to super-critical transition in single phase flow; in this transition the laminar boundary layer around the cylinder becomes turbulent and subsequently reattaches. There is some debate as to the mechanism behind this transition: Yokosawa *et al.* (1986a) attributed these changes to an increase in the free stream turbulence intensity, caused by the wakes behind the bubbles; whereas Watanabe *et al.* (1990) measured no increase in the free stream turbulence level on addition of a small amount of air bubbles. Watanabe further suggested that the change may be caused by the direct interaction of air bubbles with the boundary layer around the cylinder.

Theoretical studies of two-phase flows around bluff bodies have been largely limited to gas-solid systems due to the simplifying assumption that the only significant force acting on the particle is the drag force *e.g.* Laitone (1981); Morsi & Alexander (1972). A few studies have been published concerning bubble dispersion in flow around bodies: Lewis (1991); Nishikawa *et al.* (1991) and Cook & Harlow (1986).

An improved understanding of the phenomena of drag reduction and void redistribution occurring at low bubble concentrations is needed before engineering processes can be modelled. This study aims at establishing a foundation on which to further this understanding, by demonstrating the theoretical basis for some of these effects using a Lagrangian technique to calculate single bubble trajectories around a cylinder.

No exact equation of motion is available for a single bubble in an unsteady, non-uniform flow field, however various approximate equations of motion exist; their validity is discussed and one is selected. This generalised equation of motion for a single bubble has then been combined with potential flow solutions for flow around a cylinder to calculate bubble trajectories. These trajectories demonstrate

University of Cambridge, U.K.

the important parameters affecting the bubble motion and show some of the phenomena observed in experiments. The differences between these trajectories and those calculated using an alternative equation of motion extrapolated from low Reynolds number theory are also discussed. Direct comparison of the calculated and experimental single bubble trajectories is difficult due to the non-deterministic nature of real trajectories, however void fraction distributions can be calculated from a number of trajectories, enabling simple comparison with the previously reported experimental results of Inoue *et al.* (1986).

EQUATION OF MOTION

A key factor in any theoretical study into particle† dynamics is the formulation of an equation of motion for the particle. One of the first general equations for a particle was proposed by Basset (1888), Boussinesq (1903) and Oseen (1927) who developed an equation, using Stokes' creeping flow approximation, for a rigid sphere undergoing an arbitrary rectilinear acceleration in a stationary fluid. The equation they derived has become known as the BBO equation and analytical solutions to this equation are limited to spheres accelerating from rest in a stagnant fluid (Clift *et al.*, 1978). Similar equations have also been derived for fluid spheres *e.g.* Sy *et al.* (1971) and Morrison & Stewart (1976).

The BBO equation has subsequently been modified by numerous authors in an attempt to apply it to spherical particles in non-uniform moving fluids and in particular turbulent flow fields *e.g.* Tchen (1947), Corrsin and Lumley (1956). More recently Maxey & Riley (1983) rederived the equation of motion using a more consistent approach which allowed for the effect of flow field velocity curvature. This equation is now widely used in the modelling of small particles *e.g.* Lazaro & Lasheras (1989) and Maxey (1987).

The problem with all these solutions is that they are restricted to very low particle Reynolds number, *Re* by the creeping flow approximation. The study of larger particles has been limited, and often relies on *ad hoc* extrapolation of the low particle Reynolds number theory (Odar & Hamilton, 1964) or simplifications such as considering only the drag term (Morsi & Alexander, 1972).

Odar & Hamilton (1964) studied the simple harmonic motion of a rigid sphere at higher particle Reynolds numbers. They proposed that the BBO equation could be empirically extended to these high particle Reynolds numbers by using the two coefficients C_A and C_H to adjust the unsteady terms in the BBO equation. These two coefficients C_A and C_H were functions of the sphere acceleration, and empirical relations for them were found experimentally. The Stokes' drag was also replaced by the drag from the standard drag curve giving the following equation for the force on a particle moving at velocity V,

$$-F = \frac{1}{8} C_D \rho_l \pi d^2 V^2 + C_A V_b \rho_l \frac{dV}{dt} + \frac{C_H}{4} d^2 \sqrt{\pi \mu \rho_l} \int_{-\infty}^{t} \frac{\frac{dV}{dt}}{\sqrt{t-t'}} dt' \qquad (1)$$

where C_A and C_H are functions of the dimensionless acceleration $Ac = V^2/(d\,|dV/dt|)$:

$$C_A = 1.05 - \frac{0.066}{Ac^2 + 0.12} \qquad (2)$$

$$C_H = 2.88 + \frac{3.12}{(Ac + 1.0)^3} \qquad (3)$$

Dadi *et al.* (1991) and Sun & Faeth (1986) extended Odar & Hamilton's equation (1) to arbitrary non-uniform flow and obtained an expression similar to Tchen (1947), except with adjustable coefficients C_A and C_H :

$$-\boldsymbol{F} = -\rho_l V_b \left(\frac{D_l \boldsymbol{U}}{Dt}\right) + C_A \rho_l V_b \left(\frac{D_b}{Dt}(\boldsymbol{V} - \boldsymbol{U})\right) + \frac{3}{4} \frac{\rho_l V_b C_D (\boldsymbol{V} - \boldsymbol{U})\,|\boldsymbol{V} - \boldsymbol{U}|}{d} + \frac{C_H}{4} d^2 \sqrt{\pi \mu \rho_l} \int_{-\infty}^{t} \frac{\frac{D_b}{Dt}(\boldsymbol{V} - \boldsymbol{U})}{\sqrt{t-t'}} dt' + V_b(\rho_b - \rho_l)\boldsymbol{g} \qquad (4)$$

The right-hand side of the equation is made up of terms representing force contributions from: the pressure gradient of the undisturbed flow; the added mass of fluid travelling with the particle; the drag force; the history force; and the buoyancy force. The total derivatives are defined as the derivative following, (*i*) the motion of the bubble:

$$\frac{D_b}{Dt} = \frac{\partial}{\partial t} + (\boldsymbol{V}.\nabla) \qquad (5)$$

(*ii*) the motion of the fluid:

$$\frac{D_l}{Dt} = \frac{\partial}{\partial t} + (\boldsymbol{U}.\nabla) \qquad (6)$$

An alternative equation for particles with higher Reynolds numbers has more recently been derived by Thomas *et al.* (1983), Auton (1983), Auton (1987) and Auton *et al.* (1988). They show that the interfacial force on a sphere in an inviscid, rotational, unsteady, non-uniform flow assuming that the radius of the sphere is small compared with the scale over which the velocity gradients (strain rate) vary is given by:

† Note: the word particle is used to refer to the dispersed phase in general, *i.e.* gas bubbles, liquid drops or solid particles.

$$-F_i = -\rho_l V_b \frac{D_l \boldsymbol{U}}{Dt} + \rho_l V_b C_m \left(\frac{D_b \boldsymbol{V}}{Dt} - \frac{D_l \boldsymbol{U}}{Dt} \right)$$

$$+ \rho_l V_b C_L (\boldsymbol{V} - \boldsymbol{U}) \times \underline{\omega} \qquad (7)$$

It is additionally assumed that the change in the ambient velocity field over the scale of the sphere is small compared to the slip velocity, $|\boldsymbol{V} - \boldsymbol{U}|$. The fluid has vorticity vector $\underline{\omega}$ and the constants C_m and C_L are the added mass and lift coefficients, respectively.

Thomas *et al.* (1983) suggested that the viscous drag could simply be added to this inviscid force and including a gravitational body force, the overall expression for the force on a sphere becomes:

$$-F = -\rho_l V_b \frac{D_l \boldsymbol{U}}{Dt} + \rho_l V_b C_m \left(\frac{D_b \boldsymbol{V}}{Dt} - \frac{D_l \boldsymbol{U}}{Dt} \right)$$

$$+ \rho_l V_b C_L (\boldsymbol{V} - \boldsymbol{U}) \times \underline{\omega}$$

$$+ \frac{3}{4} \frac{\rho_l V_b C_D (\boldsymbol{V} - \boldsymbol{U}) |\boldsymbol{V} - \boldsymbol{U}|}{d} + V_b (\rho_b - \rho_l) g \qquad (8)$$

This equation has subsequently been used for modelling bubbles in vortices (Thomas *et al.*, 1983) and mixing layers (Hunt *et al.*, 1988). The right-hand side of the equation is made up from uncoupled force contributions from: the pressure gradient of the undisturbed flow; the acceleration of added mass of fluid travelling with the body; the lift due to the vorticity; the drag force; and the buoyancy force. Comparing the terms in this equation (8) with those in equation (4), three significant differences are noted: (*i*) a lift term was derived as the method allowed for velocity shear; (*ii*) no history term was derived as this is a viscous effect; (*iii*) the formulation of the derivatives in the added mass term differs. These differences are a consequence of the contrasting approaches taken in the derivation of the equations. The adapted Odar & Hamilton equation relies on extrapolation of low Reynolds theory and the Auton equation uses extrapolation of inviscid theory. It should also be noted that whilst the Auton equation is based on an inviscid solution for a non-uniform flow, the adapted Odar and Hamilton equation is further extended from the case for a stagnant fluid. It is this *ad hoc* extension that leads to the difference in the formulation of the derivatives in the added mass term. It is now generally agreed that the formulation derived by Auton is correct (Bayly, 1993).

In the current study, bubbles of diameter 0.5 - 5 mm have been used, giving particle Reynolds numbers in the range 33 - 2000, based on terminal velocities. Hence two equations of motion are suitable: the adapted Odar & Hamilton equation (4) and the Auton equation (8). Auton's equation (8) was selected for the theoretical investigation for the following reasons: (*i*) it has been implemented successfully in several cases including bubbles in vortices; (*ii*) it does not rely on *ad hoc* extrapolation of low particle Reynolds number theory; (*iii*) the assumptions made in its derivation are consistent and acceptable for high particle Reynolds numbers; (*iv*) it contains the correct formulation of the added mass term, which allows for acceleration with the fluid; (*v*) the empirical relations for Odar & Hamilton's adjustable coefficients C_A and C_H have only been formulated for particle Reynolds numbers, $Re < 80$.

However for bubbles with radii at the lower end of the range studied, the assumptions inherent in the derivation of the Auton equation may be compromised. It is for these bubbles, that can be considered rigid spheres, that the adapted Odar & Hamilton equation is perhaps the preferred equation. The effect of choice of equation on bubble trajectories is considered in the theoretical investigation by comparing the trajectories calculated with each equation.

Two questions must be considered before these higher Reynolds number equations are to be applied to bubbles rather than solid particles. Does the mobile surface of a bubble affect the validity of the equations and are the bubbles sufficiently spherical to be described by these equations? In the case of air bubbles in contaminated water the bubbles can be considered rigid and spherical for equivalent diameters up to 1.5 mm, but are ellipsoidal in the 1.5 – 5 mm range. The adapted Odar & Hamilton equation (4) depends on a rigid surface for the form of the Basset history integral to be correct, however Bayly (1993) shows that the magnitude of the Basset history force is insignificant compared to the other terms in the equation. In the case of Auton's equation (8) the nature of the bubble surface is less important as the derivation is based on inviscid theory.

Although the bubbles are ellipsoidal the same equations can still be used to describe its motion, if the added mass and drag coefficients are formulated correctly. In the extreme case they could be placed in tensorial form. However, in a turbulent flow the drag coefficient of the bubbles, less than 5 mm in diameter, can be approximated by the standard drag curve (Baker & Chao, 1965). The added mass coefficient can also be approximated to that of a sphere, *i.e.* $C_m = 1/2$.

THEORETICAL INVESTIGATION

Auton's Lagrangian equation of motion was combined with simple potential flow solutions for flow around a cylinder to calculate bubble trajectories. Void fraction distributions were then calculated by statistically combining the results from a large number of trajectories.

Equation of motion

Applying the equation (8) to a bubble in a potential flow gives:

$$\left(\frac{\rho_b}{\rho_l} + C_m \right) \frac{D_b \boldsymbol{V}}{Dt} = (1 + C_m) \frac{D_l \boldsymbol{U}}{Dt}$$

$$- \frac{3}{4} \frac{C_D (\boldsymbol{V} - \boldsymbol{U}) |\boldsymbol{V} - \boldsymbol{U}|}{d} + \left(\frac{\rho_b}{\rho_l} - 1 \right) g \qquad (9)$$

The right-hand side of the equation is made up from a convective term, a drag term and a gravitational term. There is no vorticity lift term as by definition a potential flow is irrotational. In our simulations the drag coefficient was calculated from a series of equations which form an approximation to the standard drag curve (Morsi & Alexander, 1972). The added mass coefficient C_m was taken as 0.5, as has been calculated for a sphere in an inviscid flow (Milne-Thomson,1962); recent numerical calculations by Rivero *et al.* (1991) suggest that this is also applicable for unsteady flows at *Re* of $O(10^2)$.

Assumptions

The equation is integrated over the potential flow field in a Lagrangian manner, using the following assumptions:

- the bubble can be considered approximately spherical, with drag coefficient given by the standard drag curve, *i.e.* the bubble is assumed to be immersed in a contaminated fluid;
- the viscous drag force can simply be added to the inviscid forces acting on the bubble;
- the radius of the sphere is small compared to the scale over which the velocity gradient or strain rate varies, *i.e.* bubbles are small compared with the cylinder diameter;
- the change in the ambient velocity field over the scale of the sphere is small compared to the slip velocity;
- bubble-bubble, bubble-boundary or bubble-fluid interactions may be ignored, these assumptions would be appropriate for small particle concentrations, less than about 5 - 10 %;
- the effects of turbulence are negligible and deterministic trajectories may be calculated.
- bubble coalescence and fragmentation are neglected.

Flow fields

Two potential solutions for flow around a cylinder of radius *R* were considered. The first is the well-known solution comprising uniform stream and doublet terms and gives potential (symmetric) flow around a cylinder. The second solution was chosen to simulate the effects of vortices in the wake region behind the cylinder by adding two vortex terms and two image vortex terms to the velocity potential of the first case (Milne-Thomson, 1962). This gives a wake consisting of two stationary vortices, with the vorticity confined to the vortex centres. The vortices were positioned at y co-ordinates of +/− 0.5 *R*; this constrained the vortex circulation to 5.36*UR*. This is approximately the maximum strength of a vortex shed into the wake in the sub-critical cylinder Reynolds number range (Sarpkaya & Shoaff, 1979). It should be emphasised that stationary vortices are not present at high cylinder Reynolds numbers and the flow field is simply used to demonstrate the effect of wake vortices. A subsequent study using an unsteady discrete-vortex simulation to describe the flow field is to be published in a separate paper (Bayly & Rielly, 1994). Figure 1 shows initially equally spaced streamlines for both these solutions. The first solution will be called the Simple flow field and the second, the Wake flow field.

Numerical method

To calculate the trajectory of a single bubble flowing around a cylinder, the equation of motion (9) was integrated using the potential solution for flow around a cylinder. This involved integrating a set of four simultaneous ordinary differential equations for the bubble velocity and position, as functions of time. The integration was performed using a fifth-order Runge-Kutta method with an adaptive stepsize control (Press *et al.*, 1986). Particle trajectories were started at least 10 *R* upstream from the centre of the cylinder, and were given an initial velocity equal to the uniform stream velocity *U* plus the particle terminal velocity *i.e.* the particle starts as if it had been travelling in uniform stream long enough to reach an equilibrium velocity.

The void fraction distributions were calculated using a statistical method which calculates the fraction of time a point in the flow is covered by a bubble. As the bubble trajectories are deterministic, a series of initially equally-spaced trajectories is tracked, one by one through the flow field. The total time a point is covered by a bubble is then calculated by adding the contribution of all the bubbles passing over that point. The ratio of void fraction at any point to the initial void fraction in the flow far upstream of the cylinder can then be calculated, thus giving a relative void fraction distribution throughout the flow field. To allow for continuity, bubbles which hit the cylinder are allowed to slide around, until they are convected away from the surface. It should be emphasised that as the method does not allow for bubbles affecting the liquid flow field, or for bubble-bubble interactions, it is limited to dilute bubble concentrations and the calculated relative void fraction is independent of the initial void fraction.

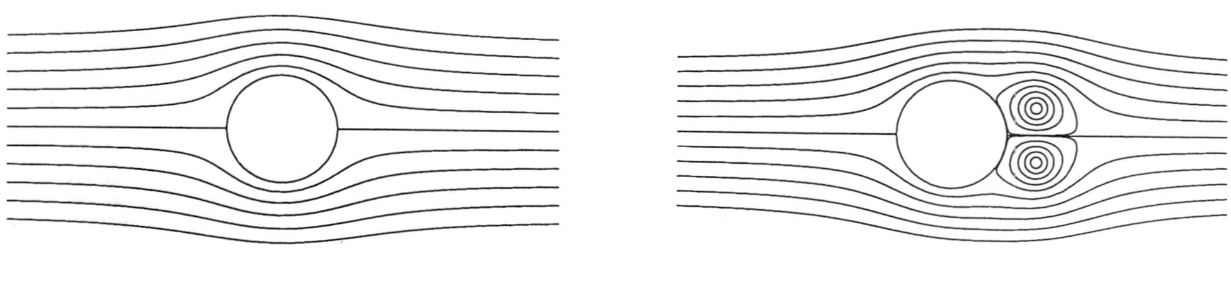

a) Simple solution. *b) Wake vortices solution.*

Figure 1 Streamlines of the flow fields used in theoretical calculations.

RESULTS AND DISCUSSION

Single bubble trajectories

Trajectories of bubbles of various diameter have been calculated using an air/water system and the Simple flow field, for a number of stream velocities and cylinder diameters. These are shown in Figures 2 - 5†. They reveal that the trajectories are dominated by two important effects: the ratio of the bubble slip velocity to the free stream velocity, and the ratio of bubble aerodynamic response time to the time scale of the flow field *i.e.* the Stokes number. For bubbles the Stokes number is equivalent to the ratio of bubble diameter to cylinder diameter (Bayly, 1993).

The bubble trajectories have generally similar shapes in these Figures, although small variations are evident. By considering the forces acting on the bubble as it moves around the cylinder these differences can be explained. As the bubble approaches the cylinder, it responds to the changing velocity gradients. In comparison to a fluid element, a bubble responds more rapidly to changes in the liquid flow as its inertia is smaller; (its effective mass is almost entirely due to the added mass of the fluid travelling with the bubble, *i.e.* half the mass of the fluid it displaces). However, the bubble slip velocity leads to differences between the derivatives following the motion of the bubble and those following the motion of the liquid, so the time period in which the bubble can respond to velocity gradients in the fluid is different from that of a fluid element. This difference is highlighted by comparison of the trajectories of bubbles in upflow and downflow situations: Figure 2 shows this for a 2.5 mm diameter air bubble in water. In downflow the bubble trajectory deviates further from the streamlines than the upflow case. As the bubble is travelling slower than the fluid in the downflow case it remains under the influence of the convective forces for a longer time period; the opposite is true for liquid upflow. These differences are highlighted in Figure 3: at higher velocities, when the slip velocity is significantly smaller than the free stream velocity, these differences become negligible and the trajectories for both upflow and downflow are similar; conversely at lower velocities the differences are increased.

The effects of bubble diameter on its trajectory are also significant (Figure 4). As the bubble becomes smaller the drag force increasingly dominates the equation of motion, and the bubbles follow the streamlines more closely as the aerodynamic response time of the bubble becomes shorter and the Stokes number decreases. The bubble size also determines the slip velocity and these two factors combine to make the bubble diameter an important variable in trajectory determination.

Decreasing the cylinder radius has a similar effect to increasing the bubble diameter (Figure 5); both increase the Stokes number and consequently this tends to increase the difference between the particle trajectory and the corresponding streamline. In many of these cases a stationary point appears behind the cylinder *i.e.* a point where the bubble almost comes to rest and a 90° change in trajectory direction can be observed. Bubble trajectories are also seen to cross in this region. This perhaps counter-intuitive effect is due to the non-linearity of the equation of motion and also partly due to the unrealistic flow pattern in this region, as the Simple flow field does not include a wake. As this potential flow solution only realistically represents the liquid velocity field around the front of a cylinder, any bubble trajectories predicted in the cylinder wake region are not physically realisable.

Calculations using the Wake flow field, which simulates a hypothetical wake region, show almost identical trajectories upstream of the cylinder, as the wake has little effect on the flow field in this area. However downstream of the cylinder the wake vortices attract bubbles that do not hit the cylinder initially. The bubbles which pass closest to the vortices are then trapped in the vortex cores as shown in Figure 6, for an upflow and a downflow. Comparing the two situations shown in these Figures shows that more bubbles are trapped in the downflow case, even though they are initially pushed further out from the cylinder. This is again due to the slip velocity of the bubbles: in downflow the bubbles move more slowly past the wake region aiding trapping; the opposite is true for upflow. Thus in upflow the rise velocity of the bubble can overcome the attraction of the vortex, when the stream velocity is low, and hence vortex strength is reduced. Smaller bubbles are also less likely to be trapped by the vortices as the drag term dominates and the bubbles follow the fluid streamlines more closely (Figure 7).

Void fraction distribution

Figures 8 - 11 show calculated relative void fraction distributions for a selection of test cases based on a 0.9 m/s water upflow around a 20 mm radius cylinder, with 4 mm diameter bubbles. The Simple and the Wake potential flow solutions predict very similar flow patterns in front of the cylinder, indicating that the addition of the wake to the potential flow solution has very little effect upstream of the cylinder. For this reason only those calculated using the Wake model are shown.

Figure 8 shows the effect of free stream velocity U on the void fraction distribution in an upflow. As the free stream velocity is increased, the void fraction distribution becomes less uniform and regions of high and low bubble concentrations are predicted. These predicted void fractions demonstrate the main phenomena observed experimentally by Inoue *et al.* (1986) (using an air/water system with mean bubble diameter of 4 mm), namely: high void fractions in the wake; a bubble free region in front of cylinder; and high void fraction streaks at the sides of the cylinder.

Both upflow and downflow simulations show vortex trapping, as demonstrated by the high void fraction regions seen leading into the centres of the vortices (marked by black circles in Figures 8 -11). The vortices collect the bubbles in the wake region, giving the high bubble concentrations that are observed experimentally.

† Note: In all the Figures flow is in the positive x direction, *i.e.* left to right, unless otherwise stated.

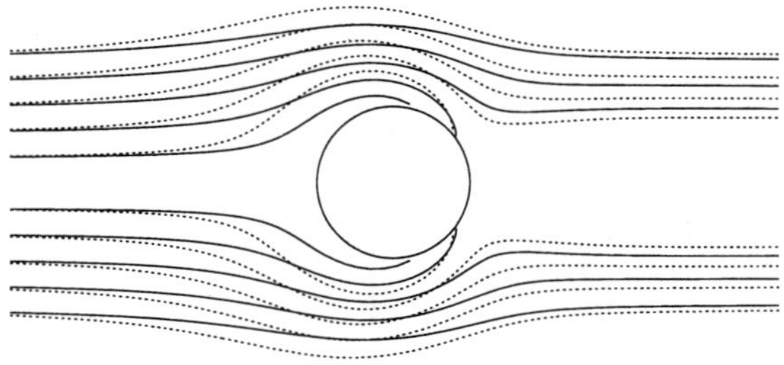

Figure 2 Comparison of trajectories in 1 m/s flow. —— upflow, ---- downflow, U = 1.0 m/s ; R = 0.03 m, d = 2.5 mm

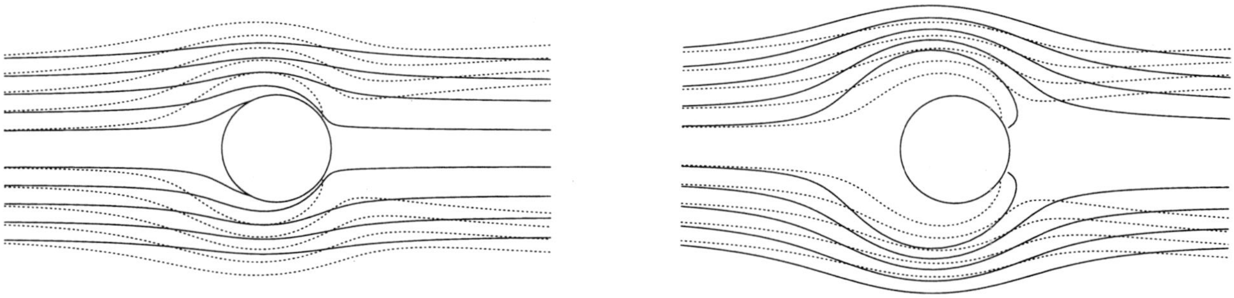

a) upflow *b) downflow*

Figure 3 The effect of velocity on bubble trajectories.—— U = 0.4 m/s, ---- U = 10 m/s, R = 0.03 m, d = 2.5 mm

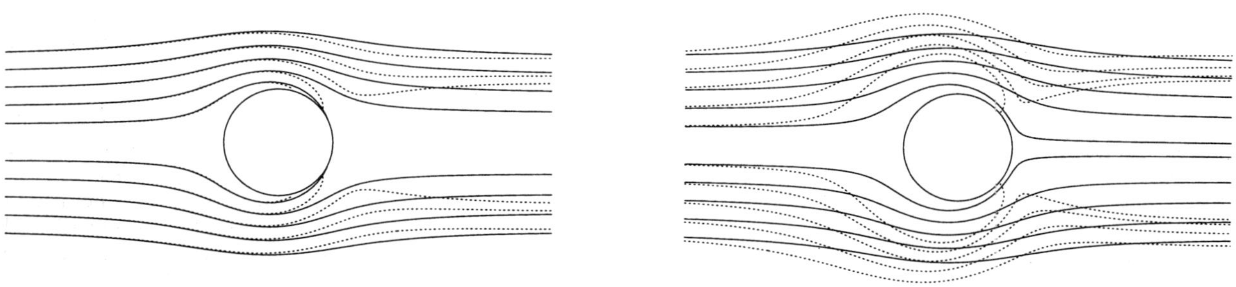

a) upflow *b) downflow*

Figure 4 The effect of bubble diameter on trajectories. —— d = 0.5 mm, ---- d = 5 mm, U = 1.0 m/s, R = 0.03 m.

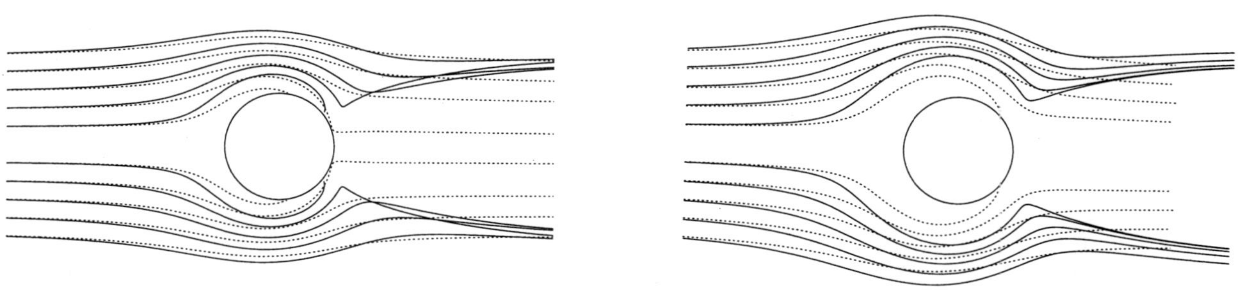

a) upflow *b) downflow*

Figure 5 The effect of cylinder radius on bubble trajectories; —— R = 0.01 m, ---- R = 0.05 m, U = 1.0 m/s, d = 2.5 mm

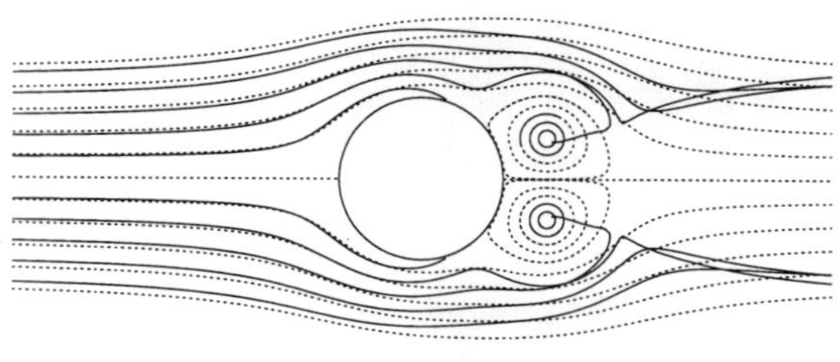

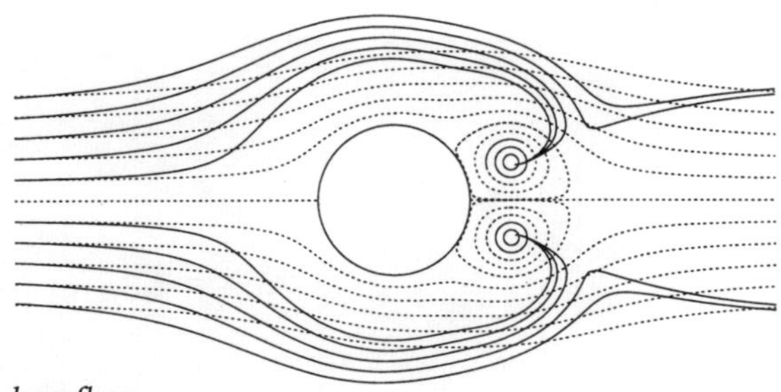

a) upflow *b) downflow*

Figure 6 Bubble trajectories using the Wake model; —— trajectory, ---- streamline, U = 0.9 m/s, R = 0.02 m, d = 4 mm

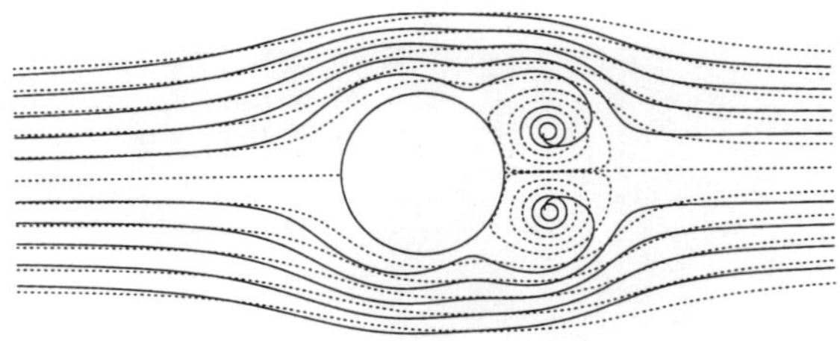

Figure 7 Bubble trajectories using Wake model; —— trajectory, ---- streamline, downflow. U = 0.9 m/s, R = 0.02 m, d = 1 mm

A bubble-free region was predicted in front of the cylinder for all cases, except for upflows at the lowest stream velocities. For the upflow cases the depth of the bubble-free region increased as the stream velocity increased, and decreased relative to cylinder radius, for larger cylinders. Table 1 compares the calculated depth of this bubble free region to that observed experimentally by Inoue *et al.* (1986). It shows that the appearance of the bubble free region can be predicted theoretically, but the method tends to overpredict the depth of the region at higher stream velocities. The size of the bubbles had little effect on the depth of the bubble free region. However, the tendency for smaller bubbles to follow the streamlines more closely led to a more uniform void fraction distribution overall (Figure 10).

In the upflow cases, bubble-boundary layer interaction is indicated for stream velocities up to 0.9 m/s, whereas at higher velocities the bubble-free region extends around the cylinder. The validity of the void fractions close to the cylinder surface is also in doubt as the cylinder surface pressure profile is only correctly predicted by potential flow theory up to 30° either side of the front stagnation point (Clift *et al.*, 1978). Having said this, the appearance of high void fraction streaks on the sides of the cylinder in some cases gives us an insight as to why they appear in experiments for the 20 mm radius cylinder (Inoue *et al.*, 1986). The bubbles which hit the cylinder are convected around the surface, causing a build up of trajectories and a high void fraction streak. This streak is less pronounced for smaller bubbles, and for smaller cylinders, as fewer bubbles hit the cylinder (Figures 10 & 11). The same trends were recorded experimentally by Inoue *et al.* (1986).

As with the trajectory predictions, large differences are seen between void distributions predicted for upflow and downflow; in downflow a larger disturbed area is seen, with a bubble-free region which extends around the sides of the cylinder (Figure 9). The depth of this region remained approximately constant at all the cylinder radii examined.

Table 1: A comparison between the theoretically calculated and experimentally observed depth of the bubble free region in front of the cylinder, for a range of stream velocities in an upflow around 10 & 20 mm radii cylinders. (Experimental data from Inoue *et al.*, 1986.)

	Cylinder Radius R = 20 mm		Cylinder Radius R =10 mm	
Stream velocity U (m/s)	Experimental depth x_{bf}/R	Theoretical depth x_{bf}/R	Experimental depth x_{bf}/R	Theoretical depth x_{bf}/R
0.45	0.0	0.0	0.0	0.0
0.90	0.1	0.1	0.15	0.15
1.90	0.25	0.4	0.3	0.6

a) U = 0.45 m/s

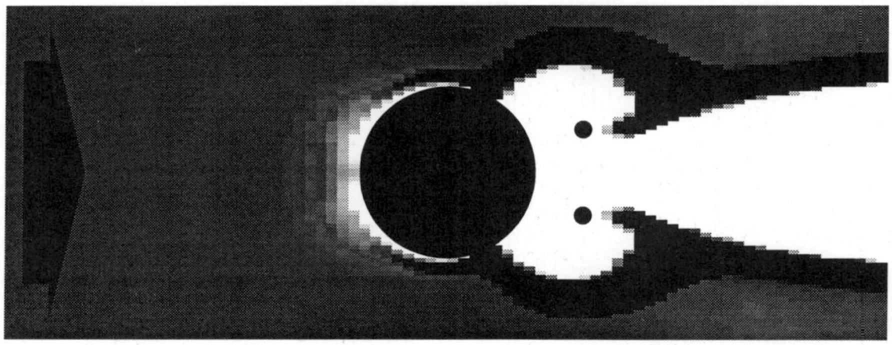

b) U = 0.9 m/s

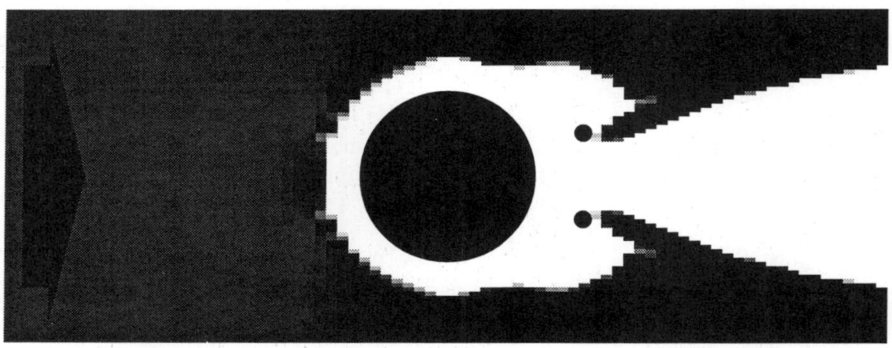

c) U = 1.9 m/s

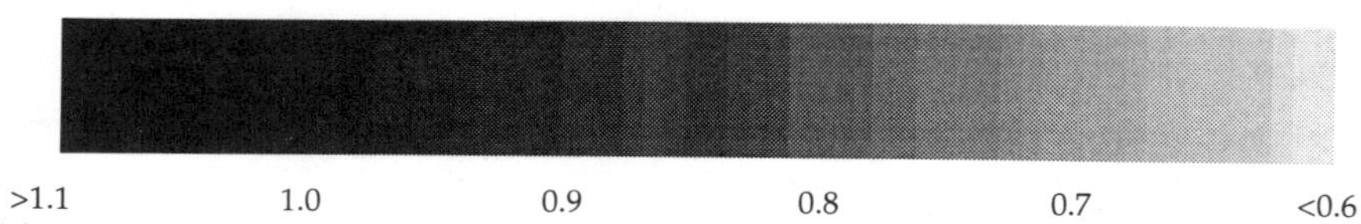

Relative void fraction scale

Figure 8 The effect of stream velocity on the air bubble void fraction distribution in water upflow around a cylinder, using the Wake flow model; R = 0.02 m, d = 4 mm.

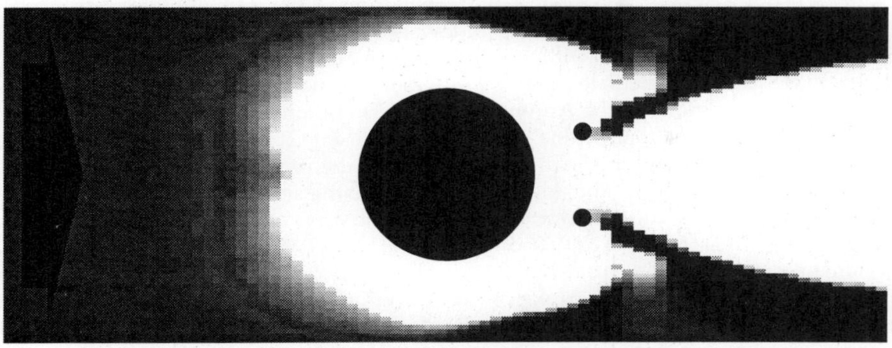

Figure 9 Theoretical prediction of air bubble void fraction distribution around a cylinder using the Wake model; downflow, U = 0.9 m/s, R = 0.02 m, d = 4 mm.

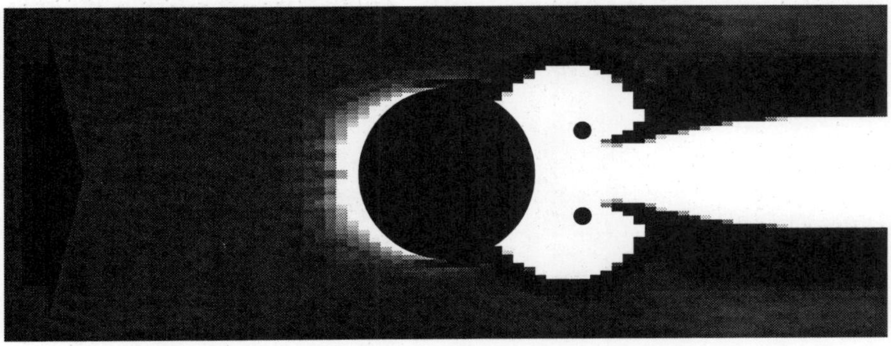

Figure 10 Theoretical prediction of air bubble void fraction distribution around a cylinder using the Wake model; upflow, U = 0.9 m/s, R = 0.02 m, d = 2 mm.

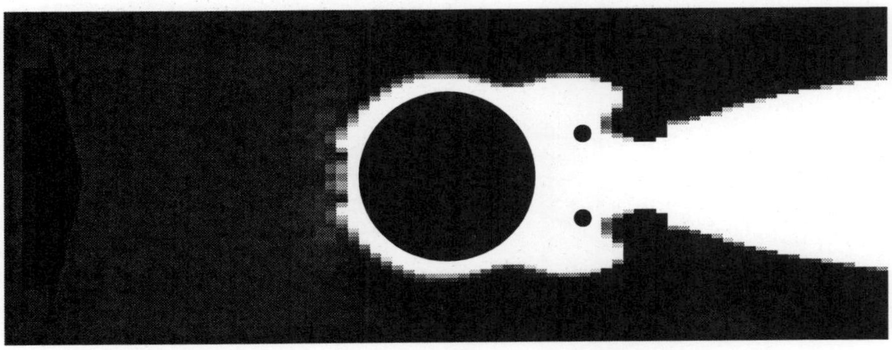

Figure 11 Theoretical prediction of air bubble void fraction distribution around a cylinder using the Wake model; upflow, U = 0.9 m/s, R = 0.01 m, d = 4 mm.

Relative void fraction scale for Figures 9 - 11.

Comparison with the adapted Odar & Hamilton equation

The adapted Odar & Hamilton type equation (4) without the Basset history force term was used to calculate single bubble trajectories under the same conditions as the Auton equation (9). The overall trends predicted were the same as those considered in the previous section for the Auton equation. In fact, for small bubble diameters and large cylinder radii, *i.e.* for low Stokes numbers, the trajectories were almost identical for both equations (Figure 12a). The largest differences between trajectories were seen at high bubble diameters and low cylinder radii, *i.e.* high Stokes number (Figure 12b), this is when the convective component in the equation is largest so the differences in the formulation and value of the added mass term are greatest. The Odar & Hamilton trajectories show less deviation from the corresponding liquid streamline as the bubble's added mass coefficient is close to 1.0 (the effective value for a fluid element), through most of its trajectory. It is this difference in the added mass coefficient that causes the most significant difference in the bubble trajectories. The difference in the formulation of the derivatives does not cause such a large deviation to the trajectories. This is demonstrated by Figure 13, which compares the trajectories produced by the two equations of motion with those produced using an adapted Odar & Hamilton equation with its added mass coefficient set to a constant 0.5. Here it can be seen that the difference caused by the use of the adjustable coefficient is large and counteracts the smaller difference caused by formulation of the derivatives.

The results suggest that the Auton equation can be used for lower bubble radii although the assumptions inherent in its derivation maybe compromised. The accuracy of the Odar & Hamilton equation at higher Reynolds is in doubt as the adjustable coefficients have not been determined in this Reynolds number range. Neither have they been obtained experimentally for high *Ac* numbers, as equations (2) & (3) were defined from low *Ac* number experiments. The physical significance of an added mass coefficient of 1.05 at high *Ac* is uncertain and disagrees with the simulations of Rivero *et al.* (1991).

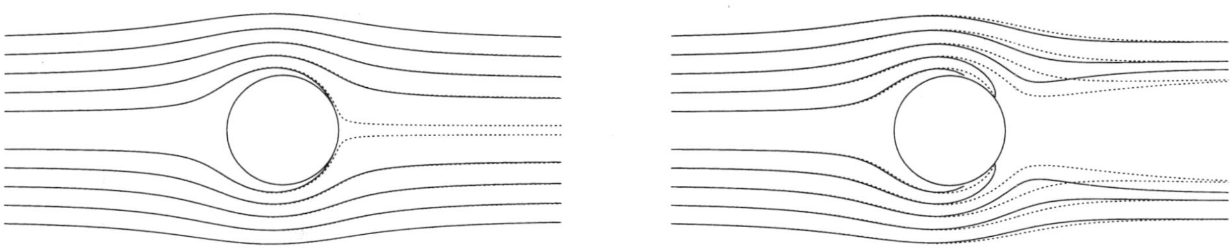

a) d = 0.5 mm *b) d = 5.0 mm*

Figure 12 Comparison of equations of motion; ---- Odar & Hamilton, —Auton, U = 1.0 m/s, R = 0.02 m, upflow

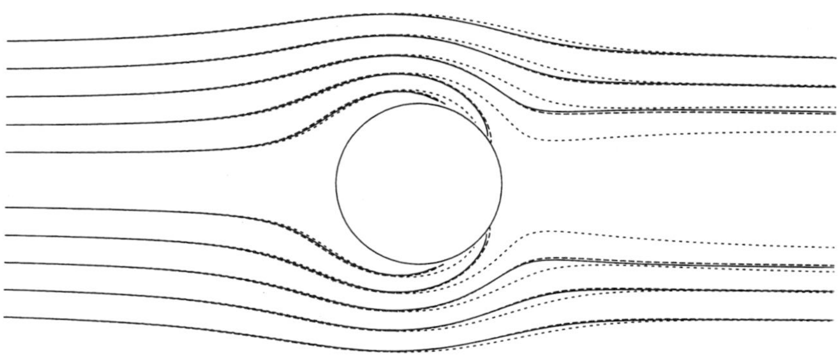

Figure 13 Effects of variable added mass coefficient on bubble trajectories; upflow.U = 1.0 m/s, R = 0.02 m, d = 2.5 mm
— Auton,
---- Odar & Hamilton with variable added mass coefficient,
– – – Odar & Hamilton with constant added mass coefficient.

CONCLUSIONS

The equation of motion suggested by Auton has been used successfully to calculate single bubble trajectories in potential flows around a cylinder. This simple model shows clearly the basic physical principles controlling the trajectories of bubbles around cylinders. The ratio of bubble to cylinder diameter and the ratio of slip velocity to stream velocity have been shown to be the most important factors affecting bubble trajectories. The latter ratio causes a significant difference between upflow and downflow situations, at low liquid stream velocities. Comparison with the adapted Odar & Hamilton equation shows identical trajectories for low diameter ratios. This suggests Auton's equation can be used at these low ratios. The differences predicted at higher diameter ratios have been shown to be mainly due to the magnitude of the variable added mass coefficient.

Void fraction distributions calculated from a large number of single bubble trajectories have demonstrated the main features observed experimentally, despite the inadequacies of the potential flow fields used. The appearance of a bubble free region in front of the cylinder could be predicted and its size could be estimated. The inclusion of a hypothetical wake in the potential flow solution demonstrated the importance of vortex trapping of bubbles. This effect is likely to dominate the bubble motion in real flows, and the effects of vortex trapping in unsteady shed wake vortices are the subject of a current investigation

Overall it can be concluded that a solid foundation has been established as a basis for more extensive research, which will lead to a wider and improved understanding of bubbly flow around bluff bodies

ACKNOWLEDGMENT

This work was supported by the Science and Engineering Research Council.

NOMENCLATURE

Roman symbols

Ac	Acceleration number ($V^2/(d\,\lvert dV/dt\rvert)$)	
C_A	Odar & Hamilton's (1964) added mass coefficient	
C_H	Odar & Hamilton's (1964) history coefficient	
C_D	Drag coefficient	
C_m	Added mass coefficient	
C_L	Lift coefficient	
d	Particle diameter	m
$\mathbf{F}$	Force acting on particle	N
g	Gravitational acceleration	m/s^2
R	Cylinder radius	m
t	Time	s
U	Stream velocity	m/s
$\mathbf{u}$	Velocity of fluid	m/s
$\mathbf{V}$	Velocity of particle	m/s
V_b	Particle volume	m^3
x	Distance from cylinder centreline in stream direction	m
x_{bf}	Depth of bubble free region in front of cylinder stagnation point	m
y	Distance from cylinder centreline perpendicular to stream direction	m

Greek symbols

μ	Continuous fluid viscosity	kg/ms
ρ_l	Continuous fluid density	kg/m^3
ρ_b	Particle density	kg/m^3
$\underline{\omega}$	Vorticity vector	

REFERENCES

Auton, T.R., 1983, Ph.D. Thesis, University of Cambridge.

Auton, T.R., 1987, *J. Fluid Mech.*, **183**, 199 - 213.

Auton, T.R., Hunt J.C.R., Prud'Homme, M., 1988, *J. Fluid Mech.*, **197**, 241-257.

Baker, J.L.L., Chao, B.T., 1965, A.I.Ch.E. J., **11**, 268 - 273.

Basset, A.B., 1888, "*A Treatise on hydrodynamics*", Vol. 2, Deighton Bell, Cambridge.

Bayly, A.E., 1993, CPGS Dissertation, University of Cambridge.

Bayly A.E., Rielly, C.D., 1994, Submitted to ASME - Cavitation and Multiphase Flow Forum, 19-23 June, Lake Tahoe, U.S.A.

Boussinesq, J., 1903, "*Theorie analitique de Chaleur*", Vol. 2, Paris, 224.

Clift, R, Grace, J.R., Weber, M.E., 1978, "*Bubbles, drops and particles.*", Academic Press.

Cook, T.L., Harlow, F.H., 1986, *Int. J. Multiphase Flow*, **12**, 35-61.

Corrsin, S., Lumley, J., 1956, *Appl. Sci. Res.*, **A6**, 114-116.

Dadi, M., Stanislas, M., Rodriguez, O., Dyment, A., 1991, *Experiments in Fluids*, **10**, 285-294.

Hunt, J.C.R., Auton, T.R., Sene, K., Thomas, N.H., Kowe, R., 1988, "*Transient Phenomena in multiphase flow*", Ed. Afgan, N.H., Hemisphere, USA, 103-126.

Inoue, A., Kozawa, Y., Yokosawa, M., Aoki, S., 1986, *Int. J. Multiphase Flow*, **12**, 149-167.

Laitone, J.A., 1981, *Trans. ASME - J. Appl. Mech.*, **48**, 465-471.

Lazaro, B.J., Lasheras, J.C., 1989, *Phys. Fluids*, **A1**, 1035-1044.

Lewis, P.D., 1991, CPGS Dissertation, Cambridge University.

Maxey, M.R., Riley, J.J., 1983, *Phys. Fluids*, **26**, 883-889.

Maxey, M.R., 1987, *Phys. Fluids*, **30**, 1915-1928.

Milne-Thomson, L.M., 1962, "*Theoretical Hydrodynamics*", MacMillan, London.

Morrison, F.A., Stewart, M.B., 1976, *Trans. ASME - J. Appl. Mech.*, **43**, 399-403.

Morsi, S.A., Alexander, A.J., 1972, *J. Fluid Mech.*, **55**, 193-208.

Nishikawa, H., Masumoto, Y., Ohash, H., 1991, *Computers & Fluids*, **19**, 453-460.

Odar, F., Hamilton, W.S., 1964, *J. Fluid Mech.*, **18**, 302-314.

Oseen, C.W., 1927, "*Hydrodynamik*", Leipzig, 132.

Press, W.H., Flannery, B.P., Teukolsky, S.A., Vetterling, W.T., 1986, "*Numerical Recipes*", C.U.P.

Rivero, M., Magnaudet, J., Fabre, J., 1991, *C. R. Acad. Sci. Paris, -Mécanique des fluides*, Série II, t. **312**, 1499-1506.

Sarpkaya, T., Shoaff, R.L., 1979, AIAA *Journal*, **17**, 1193-1200.

Sun, T-Y, Faeth, G.M., *Int. J. Multiphase Flow*,**12**, 99-114.

Sy, F., Lightfoot, E.N., 1971, *A. I. Ch. E. J.*, **17**, 177-181.

Tchen, C.M., 1947, Ph.D. Thesis, Delft University, Netherlands.

Thomas, N.H., Auton, T.R., Sene, K., Hunt J.C.R., 1983, *BHRA Int. Conf. Physical Modelling Multiphase Flow*, Coventry, England.

Van't Riet, K., Boom, J.M., Smith, J.M., 1976, *Trans. IChemE*, **54**, 124-131.

Watanabe, H., Ihara, A., Onuma, S., 1990, *Trans. ASME - J. Fluids Eng.*, **112**, 67-72.

Yokosawa, M., Kozawa, Y., Inoue, A., Aoki, S., 1986a, *Int. J. Multiphase Flow*, **12**, 169-184.

Yokosawa, M., Kozawa, Y., Inoue, A., Aoki, S., 1986b, *Int. J. Multiphase Flow*, **12**, 185-202.

Characterizing Fluidized-Bed Behavior by Decomposition of Chaotic Phase-Space Trajectories

J. S. Halow
U.S. Department of Energy, Morgantown Energy Technology Center, Morgantown, WV 36507

C. S. Daw
Oak Ridge National Laboratory, Oak Ridge, TN 37831

Recent applications of chaotic time-series analysis to gas fluidized beds have demonstrated that substantial information about fluidization conditions within the bed can be extracted from voidage and pressure drop data. We present a technique to characterize fluidized-bed behavior that is based on the crossings of the phase-space trajectory through the principal-component planes. Starting with either pressure drop or void fraction versus time data, time-series embedding and principal-component analysis is used to construct a phase-space trajectory for the data. This trajectory characterizes the dynamic state of the bed. The technique "decomposes" the trajectory by sorting the orbits into characteristic types for different modes of bed behavior, such as emulsion phase fluctuations, bubbling, slugging, and bubble coalescence. We present the method and an analysis of data from experiments in several fluidized beds. The overall goal of these studies is to improve the diagnostics and control of fossil-energy fluidized-bed processes.

The Morgantown Energy Technology Center has development programs in fossil energy technologies including fluidized-bed combustion, coal gasification and hot gas cleanup. Many of these systems use fluidized beds as processing reactors, heat exchangers, or other gas-solids contacting devices. Diagnosing bed behavior and controlling these processes are critical to efficient system performance. Our goal was to develop novel techniques to improve diagnosis and system control.

Recent studies [1-8] demonstrated that chaotic time-series analysis can characterize the dynamic patterns of fluidized beds by continuous measurements. These results have stimulated the hope that it may be possible to extract previously unused information about fluidized-bed hydrodynamics by employing chaotic time-series analysis of fluidized-bed signals. Such information could be used to improve the monitoring, control, and even the design of fluidized-bed reactors.

In this paper, we focus on recognizing characteristic patterns in the phase-space trajectories produced by chaotic time-series analysis of fluidized-bed pressure-drop and voidage signals. Phase-space trajectories are dynamic maps produced by time-series embedding and principal-components analysis (or singular value decomposition) [1,9]. These maps are multidimensional representations that essentially summarize all dynamic information contained in a measured time series. The analytical power of these maps is that they require no assumptions about the form of the time-series correlations (i.e., linear and nonlinear processes are equally well represented), and, at least in theory, they reflect information about all important dynamic components, even those that are not directly measured.

A major difficulty in utilizing phase-space trajectories is that their multidimensional structure makes direct visualization difficult. Also, their patterns are much more detailed than is shown by characterization with only a few parameters such as fractal dimension and entropy. Thus, we have tried to develop trajectory analysis methods that are sensitive to dynamic details while providing relatively simple output parameters, which can be used to compare fluidization states.

Our key objective has been to develop a method for "decomposing" (breaking down) phase-space trajectories into their characteristic elements. We have tried to maintain the following constraints:

- To develop an analytical formalism that would be amenable to automation and computerization, avoiding the need for direct human visualization of trajectories;

- To develop methods that utilize both multivariate statistics and time-correlation information present in phase-space trajectories; and

- To develop methods that would allow correlation between phase-space structure and specific physical phenomena.

Our first attempts at trajectory decomposition are presented in the following results. The methods are based on decomposition of a limited number of experimental time series. Our intention is not to provide a complete, self-contained set of analytical methods, but rather to introduce some useful techniques for trajectory decomposition. We hope this will stimulate further development. Other authors have explored methods of trajectory decomposition for other systems. See, for example, Jansen and Nyberg's [10] approach.

TIME-SERIES EMBEDDING AND PRINCIPAL-COMPONENTS ANALYSIS

In order to produce phase-space trajectories, time-series measurements are transformed by the time-series embedding technique [1,5]. The embedding process reconstructs a full-dimensional facsimile of the system's dynamics by creating multidimensional vectors from the measured variable, according to the prescription

$$\underline{p_t} = \{x_t, x_{t+k}, x_{t+2k}, \ldots x_{t+(m-1)k}\}$$

where k is an integral number of sampling intervals and m is the integral dimension of the embedding space. The procedure for identifying appropriate values for k and m is an important aspect of time-series embedding, but is beyond the scope of this discussion. Methods for this procedure can be found elsewhere [1,5]. We use typical values for k and m of k=1 and m=10. Our sampling rate (the reciprocal of the time interval between measurements) typically ranges between 62.5 and 200 Hz.

Each time series can thus be transformed into a sequence of points $\underline{p_t}$ that represents a map of the dynamic trajectory in m-dimensional space. When a deterministic process is operated in a steady dynamic state (i.e., the operating parameters are constant), the trajectory will tend to visit a subregion of the complete embedding space and follow an "attractor." An attractor represents the long-term trajectory of the system after initial transients have died out. Chaotic processes such as fluidized beds have attractors that tend to be complex and must be described with fractal geometry. Figure 1 shows a two-dimensional projection of a typical phase-space trajectory for fluidized-bed void-fraction data.

Another characteristic of chaotic attractors is that their trajectories are composed of an infinite number of similar but nevertheless distinct cycles or "orbits." Each orbit never repeats exactly, but families or categories of orbits with generally similar behavior do appear. Principal-components analysis provides a secondary transformation of the reconstructed trajectory, which creates a new set of orthogonal axes that align with the principal moments of the attractor [5]. In effect, the attractor is rotated to obtain the best possible views when it is projected into lower-dimensional spaces. Our experience with fluidized-bed data indicates that the new coordinates correspond approximately to the n-1 order derivatives of the time series; i.e., principal component 1 is approximately proportional to the time-series variable locally averaged over the embedding dimension; principal component 2 is approximately proportional to the first derivative of the time-series variable; principal component 3 is approximately proportional to the second derivative; and so on. The absolute numeric values depend on the reconstruction method. The units of principal components, however, all remain equal to the units of the original time series, because the components are constructed from linear combinations of the original measurements. We often make principal components dimensionless by normalizing them with the time-series standard deviation.

We find that most of the relevant information for fluidized beds occurs in the first three to five components. We typically attempted the following procedures only with two- or three-dimensional projections of the trajectories.

PRINCIPAL-COMPONENT-PLANE CROSSING ANALYSIS

Characteristic orbits in the phase-space trajectory of a system can be classified into groups or families that approximately represent the various dynamic states of the system. At any specific location in a bubbling or slugging bed, the local fluidization condition will alternate between an "emulsion phase" and a condition in which a void is moving through the sensing volume. In a bubbling bed, voids of various sizes and velocities will be present, depending on the time history of void formation at the grid and coalescence up to the point where measurements are made. In slugging beds, the diameter of the slugs is relatively constant; however, their length, velocity, and the manner in which solids flow around the slugs can vary, depending on the time history of the void formation at the grid and coalescence.

It is difficult to select a criterion for groups that differentiates orbits based on general characteristic features and yet ignores less important variations in the orbits within a group. We are developing a method that appears to have the desired characteristics for use with fluidized-bed data. We are calling this method principal-component-plane crossing analysis or crossing analysis for short.

Constructing the Crossing String

We assume that essential features of the trajectory can be represented by the temporal sequence in which the trajectory crosses a predefined set of orthogonal planes. The crossings for each plane represent a Poincare section of the trajectory. The predefined planes were initially chosen to be principal-component planes or planes parallel to them. The sequence of the trajectory crossings of these planes is called the crossing string.

The trajectory may cross each plane in different ways. For each plane, the crossings can be in any one of four quadrants and in one of two directions within each quadrant. Therefore, 4 x 2 possible crossings exist per plane, and n x 4 x 2 crossing types exist per set of n planes. There are 16 crossings in a two-dimensional set and 24 in a three-dimensional set. Figure 2 provides a nomenclature for the possible crossings in a three-dimensional set of orthogonal planes, and Figure 3 describes a two-dimensional crossing set. The two-dimensional set is a subset of the three-dimensional crossing definitions. Letters A through X are assigned to the various crossings. This arbitrary assignment serves to uniquely identify the various crossing events. A crossing string consists of a sequence of these letters.

The PC1 versus PC2 projection of the phase-space trajectories for fluidized beds may appear as either "left-sided" in which the left end of the trajectory appears open or "right-sided" where the right side appears open. Left-sided trajectories show a counter-clockwise rotation or spiralling around the PC1 axis, and right-sided trajectories rotate or spiral clockwise around the PC1 axis. Some of the crossings defined in Figure 2 are only associated with left-sided trajectories and others only with right-sided trajectories.

Pressure drop data yield left-sided trajectories when the bed is smoothly bubbling; right-sided trajectories are characteristic of slugging. In both types of trajectory, the left and right (negative and positive) fluctuations of PC1 are a reflection of negative and positive fluctuations in pressure drop. (PC1 is approximately proportional to the pressure drop itself.) Negative pressure fluctuations occur when bubbles passing near or between pressure taps decrease the local solids concentration, and positive pressure peaks occur after bubbles leave and solids flow back in to fill the space. The left-sided appearance of bubbling-bed trajectories occurs because the upward bubble motion is more abrupt than the return solids flow, causing negative pressure peaks to be smaller than positive peaks (and the PC1 deviations to be larger on the left than the right). Conversely, the solids falling through and around slugs are much more violent than the rise of the slug itself, causing positive pressure and PC1 peaks to be larger than the negative. We have observed that left-sided trajectories also show a counter-clockwise rotation or spiralling around the PC1 axis, while right-sided trajectories rotate or spiral clockwise around the PC1 axis.

Once the dimensionality is chosen, a crossing string or crossing sequence can be defined. Each crossing of a principal component plane is defined by a change in sign of that principal component. The entire sequence can be visualized as a string of letters that represent the temporal sequence of crossings. For more detailed analysis, the time of each crossing should be retained.

Variations in How the Crossing String is Defined

Crossing-plane analysis can be carried out in two or more dimensions. Higher dimensional methods generate a more detailed representation of the trajectory and identify more sub-groups of orbital patterns. However, relatively low selectivity is better, for example, to separate emulsion phase orbits from those caused by bubbles. In this case, a two-dimensional analysis is sufficient since it differentiates orbits that cross the PC1=0 plane from those that do not.

If the orbits associated with voids are to be split into subcategories, then a more discriminating method is needed. One such method considers, in addition to the crossing of the PC1 and PC2 planes, whether the crossing was in front of or behind the PC3=0 plane, which is readily determined by the sign of PC3. One variation of this method is to consider the sign of PC3 only for the crossings of the PC1=0 plane. We refer to this as a 2-D1P method. Similarly, a method can be defined that considers the sign of PC3 when the PC2=0 plane is crossed, referred to as a 2-D2P method. In the same manner, a 2-D1P2P method considers the sign of PC3 at crossings of both PC1 and PC2. All of these methods are subsets of the 3-D crossing set.

Another useful variation is to translate the origin of the crossing-plane set along one or more of the axes so that the set includes or excludes certain orbits. In the case of void fraction measurements, translating the origin along the PC1 axis is equivalent to setting a minimum peak size in the average void fraction that is considered a bubble. This variation often provides a clearer delineation of orbits. In the same manner, shifting the origin along the PC1 axis for differential and overall pressure drop measurements is equivalent to defining the minimum pressure fluctuation that is considered a bubble.

Crossing String Patterns

The most obvious feature of trajectories from fluidized-bed data is the loop nature of the overall orbits. This is true for void fraction, differential pressure drop, and overall pressure drop data. We define individual orbits for these loops as segments of the crossing string that begin and end with the same crossings, typically the PC2 crossing plane. This is again somewhat arbitrary, but is reasonable for dense phase fluidization. Orbits are defined with the initial and final crossing as crossings of the PC2=0 plane at values of PC1<0. In addition, we stipulate that the crossings are with PC2 and go from negative to positive or in the upward direction for right-sided trajectories. In the case of 2-D and 2-D1P analyses with right-handed trajectories, an M crossing occurs. In the case of 2-D2P, 2-D1P2P, and 3-D analyses, either an M or an O crossing occurs.

Considering the two-dimensional crossing plane set, a loop can be present in the region PC1<0, where the trajectory crosses back and forth across the PC2=0 plane. Common crossing patterns are illustrated in Figure 4, which shows this type of loop as an "MNM" type. A second type of loop occurs with the following sequence of crossings:

"M crossing" or PC2=0 at PC1 < 0
"A crossing" or PC1=0 at PC2 > 0
"J crossing" or PC2=0 at PC1 > 0
"F crossing" or PC1=0 at PC2 < 0
"M crossing" or PC2=0 at PC1 < 0

This is the large loop sequence that is typical of a bubble in the bed.

In addition to these two types, a back and forth crossing of the PC2 plane at values of PC1>0 leads to "JIJ" sequences within the MAJFM loops. Loop patterns of the following types will then be defined:

MNM
MAJFM
MAJIJFM
MAJIJIJFM
MAJIJIJIJFM
.
.
.
MA{n[JI]J}FM

For the 2-D1P crossing, the variations are:

- - MNM

MAJFM	MCJFM	MAJHM	MCJHM
MAJIJFM	MCJIJFM	MAJIJHM	MCJIJHM
MAJIJIJFM	MCJIJIJFM	MAJIJIJHM	MCJIJIJHM
.	.	.	.
.	.	.	.
MA{n[JI]J}FM	MC{n[JI]J}FM	MA{n[JI]J}HM	MC{n[JI]J}HM

Other variations in two-dimensional patterns can be defined in a similar manner. The number of potential loop types grows rapidly in three or more dimensional crossing strings. In three dimensions, the sign of the PC3 component at the times of the PC1 and PC2 crossing adds two possibilities for each crossing, leading to eight potential variations for the basic MNM sequence and 32 potential variations in the MAJFM sequence. In addition, an even number of crossings of the PC3 can occur, which leads to virtually unlimited variations in the MAJFM pattern. With the increase in the complexity of the loop description comes a much higher degree of discrimination. Three-dimensional loops with identical crossing string patterns generally are very similar.

From a practical point of view, two-dimensional loop patterns and simple extensions of the two-dimensional description are usually sufficient to decompose a trajectory into its major features.

Orbit Characteristics

Fluidized beds exhibit various local dynamic states, and the average orbital period or the percent of time spent in each of the states varies with changes in operating conditions. Patterns for each of these states have certain characteristics, including the time required to traverse the orbit, and maximum and minimum values of the orbit in each of the principal component directions. For the various patterns present in a particular trajectory, we can calculate average orbital periods, frequency of occurrence, and the percentage of time that the system spends in each orbital type.

A bed at incipient fluidization will exhibit emulsion phase fluctuations but no bubbles. Such a bed would have a phase-space trajectory with many MNM orbits but no MAJFM or MAJIJFM orbits. At higher superficial velocities, voids yield MAJFM orbits that have orbital periods related to the length and rise velocities of the voids. The percent of time spent in this dynamic state increases at the expense of the MNM state. At higher velocities in smaller beds where slugging occurs, MAJIJFM orbits and orbits with more than one [JIJ] loop in it are typical. These orbits usually represent long slugs.

MAJIJFM type orbits can, however, also occur in de-fluidized beds where the trajectory wanders repeatedly around the PC1 axis. Also, some void fraction trajectories show this type of pattern when bubbles are coalescing.

TESTING THE METHOD

Decomposition Examples

We discuss the average void fraction in a plane and overall pressure-drop data from three experiments in a 0.15-m (6-in) diameter bed. The experimental apparatus and procedure have been previously described [11].

Briefly, the bed was a 0.15-m diameter bed fluidized by air at ambient conditions. The bed was fluidized at 0.26 to 0.6 m/s or at Uo/Umf ratios of from 1.35 to 3.18. A porous plate grid was used to introduce the air. The material was an acrylic powder with particles that were irregularly shaped, 704 μm in average size, and with a minimum fluidization velocity of 0.19 m/s. The settled bed height in the experiments was 0.8 m. Overall pressure drop was determined using a pressure tap mounted at the bed wall 0.025 m above the grid.

A capacitance imaging system was used to measure the void fraction in a region from 0.47 to 0.57 m above the grid. This system generated detailed three-dimensional time-varying images of the voids, which were concurrent with the overall pressure drop. These images were averaged over the bed cross-section to give average void fractions versus time data for the 0.025-m thickness of the imaging slice. These images helped associate the physical phenomena in the bed with characteristics of the phase-space trajectories generated from the average void fraction and pressure drop data.

Bubble data for the experiment with a superficial velocity of 0.60 m/s obtained from the bed imaging included a bubble frequency of 1.1, average frontal diameter of 0.13 m, average length of 0.33 m, and average bubble velocity of 0.84 m/s. These parameters were determined for each of the 20 voids observed in the experiment. Seventeen of these voids had lengths significantly longer than their diameters. The bed was best described as slugging. Table 1 gives the measured characteristics of each void as it passed through the imaging zone. Void fraction data from this experiment had an estimated correlation dimension of 2.8 and a Kolmorogov entropy of 2.5.

Figure 5 shows the PC2 versus PC1 projection of the phase-space trajectory for the void-fraction data and its decomposition, calculated by crossing analysis. The full trajectory is shown in the top center. A two-dimensional decomposition split the trajectory into the MNM portion shown on the upper left and the MAJFM and MAJIJFM orbits shown below the full trajectory. Capacitance images showed that the MNM fragment represents small fluctuations in the emulsion phase between voids. The detailed images also showed that both the MAJFM and the MAJIJFM fragments represent orbits that are characteristic of voids passing through the imaging zone.

The difference between the MAJFM and MAJIJFM orbits is the small loop around the PC1 axis in the MAJIJFM orbits. Detailed imaging showed that this represents a variation in the cross-sectional area of the slug that is caused by irregular solids flow around the slug. A 2-D1P decomposition split out the MCJHM trajectory, which corresponded (again shown by the detailed imaging) to coalescence of two bubbles in the imaging region. Table 2 lists the parameters for both decompositions.

Some of the characteristics of the MAJFM and MAJIJFM orbits of the void-fraction trajectory are related to physical parameters in the bed. The maximum of PC1 of each orbit is related to the diameter of the void, based on its maximum cross-section. Figure 6 illustrates the correlation of void diameter versus maximum PC1 value in each of the MAJFM and MAJIJFM grouped orbits for several superficial velocities.

If each orbit represents the traversing of the imaging zone by a void, the orbital period should be equal to the void length divided by its velocity. Figure 7 shows a plot of the orbital period for each void versus the length divided by velocity. A general trend is evident, although there is some scatter. Also, the half orbit, or the time between the crossings of the PC1=0 plane, is more nearly equal to the length divided by velocity. The full orbit apparently contains the void-fraction disturbances ahead of and behind the voids. Data from two experiments are included in this figure.

Figure 8 represents the phase-space trajectory of the overall bed pressure drop and its decomposition. This corresponds to the void-fraction trajectory and decomposition illustrated in Figure 5. A typical orbit would start on the left side of the trajectory with a sequence of loops around the PC1 axis, with each loop ending further to the right than the preceding loop. A sequence of dips usually follows. Finally, the trajectory moves to its extreme right position and loops back to the beginning. Overall pressure drop shows the dynamics occurring at the bed surface as well as what is occurring near the pressure tap. Comparison of the overall pressure signals with the simultaneous imaging of the midregion of the bed implies that the initial loops and dips represent voids passing the lower tap (these appear as voids in the imaging after some time lag), and the large final loop represents voids reaching the bed surface and dissipating (these appear as voids in the imaging at an earlier time than the pressure signals). The decomposition of the pressure drop trajectory in Figure 8 shows that voids at the surface are the MAJFM trajectories. The grid level voids are represented by the orbits in the MNM section. The PC1 crossing plane was shifted to a value of PC1=1500 to provide a clean separation of these two trajectories.

The correspondence of the large loops in the overall pressure drop trajectory to voids at the bed surface is illustrated in Figure 9. This figure compares the time occurrence of the maximums in PC1 values of the MAJFM and MAJIJFM orbits for void-fraction versus pressure-drop data. Maximum PC1 values are plotted against the times they occurred in the experiment. The open circles are the PC1 maximum values from the void-fraction trajectory, and the filled circles are the overall pressure drop divided by 100. The observation times for voids in the imaging zone correspond to the void-fraction maximums. A one-to-one correspondence occurs, with pressure maximums lagging behind the void-fraction maximums.

The size of the PC1 maximum in the MAJFM pressure loops is related to the maximum expansion of the bed, which occurs just before the void reaches the surface. Figure 10 shows these maximums versus the volume of voids that reach the surface. Void volume should approximate expansion in the current system. A coincidental velocity effect appears to be proportional to the square of the ratio of superficial velocity to minimum fluidization velocity. At the highest velocity, the PC1 maximums "peak out" because the bed has expanded into a conical disengaging section that is 0.96 m above the grid.

Fluidization Regimes

Fluidization in a particular system changes with increasing superficial velocity. As velocity increases, voids become more frequent and they increase in size until slugging occurs. In some systems, fluidization transitions to a turbulent regime and finally to a transported bed.

At the University of Tennessee, differential pressure data was acquired on a well characterized fluidized bed over a range of fluidization regimes. The experimental setup, data acquisition methods, and chaotic analysis of the data have been reported elsewhere [5]. One set of experiments used a distribution of fine steel spheres that had an average diameter of 175 μm. The minimum fluidization velocity was 1.00 m/s. The bed was a 0.102-m diameter bed operated at ambient conditions. Data was acquired by pressure taps set 0.127 m apart, with the lower tap located 0.23 m above the distributor.

Figure 11 shows the differential pressure-drop phase-space trajectories for three experiments at varying the superficial velocity. At a velocity of 1.51 x Umf the bed is in a bubbling regime; at 2.4 x Umf the bed is slugging; and at 4.47 x Umf, the bed is turbulent. A total of 10,000 data points were collected in each 50-second experiment.

Differential pressure measurements show the highest pressure drop when only the emulsion phase is present. The pressure drop decreases when a void is in the bed between the taps. The large orbits in these phase-space trajectories start at the higher values of PC1, drop to lower values of PC1 as the void passes the lower tap, and then rise again as the void passes the top tap. This behavior for differential pressure orbits was confirmed by comparing videos made of the University of Tennessee bed with the actual orbits. The behavior was independently confirmed in the METC system where images of the bed were recorded while in-bed differential pressure probes recorded the pressure drop in the same region.

Crossing pattern orbits caused by bubbles passing the probes begin and end with an I or J crossing; an M or N crossing is the end of void-fraction and overall pressure-drop trajectories. To define comparable patterns (ones that begin and end with an M crossing), mirror image or reverse trajectories were constructed for the local pressure-drop data and were then decomposed. This is equivalent to defining the imbedded variable as -1 times the pressure drop, rather than the direct pressure drop. The MNM patterns in these reverse trajectories represent the emulsion phase fluctuations between voids, and the MAJFM and MAJIJFM patterns represent voids passing the taps. This behavior was again confirmed by video tapes of the University of Tennessee bed and images of the METC bed.

Figure 12 shows the percent of time spent in each of the principal types of orbits for the three velocities. The emulsion phase orbit MNM decreases with increasing velocity. The typical bubbling pattern "MAJFM" also decreases with increasing velocity. More complex MAJIJFM and higher [JIJ] patterns increase with increasing velocity. Figure 13 shows the probability of each orbital time occurrence versus orbital time. MNM orbits are typically 0.1 to 0.2 s, and MAJFM and MAJIJFM orbits associated with voids are about 0.4 to 0.6 s. Figures 13A and 13B show that the periods of MAJFM orbits increase, suggesting the longer or slower voids that would be expected in a slugging bed. Significantly, the distribution of void orbital times becomes wide in a turbulent bed.

CONCLUSIONS

We present a method for decomposition of phase-space trajectories for void fraction and overall and local pressure drops. This "principal-component-plane crossing analysis" method characterizes a trajectory as a temporal sequence of the crossings of an orthogonal set of phase-space planes. We show that the method can separate the orbits caused by voids from the orbits caused by small-scale fluctuations in the emulsion phase.

The phase-space trajectories for void fraction data yield the following information:

- Void size is proportional to the maximum of PC1 of the MAJFM and MAJIJFM orbits.
- The half orbital period is approximately equal to the length of the void divided by its velocity.
- Large slugs with irregular cross-sections are represented by MAJIJFM orbits in a two-dimensional decomposition where the JIJ loop is on the PC2>0 leg of the orbit and near the maximum value of PC1.
- Coalescing voids are represented by orbits with either loops or dips on the return

(PC2<0) side of the orbit. These orbits may appear as MAJIJFM orbits in a two-dimensional decomposition. These loops can be distinguished from other phenomena by higher order decompositions.

The phase-space trajectories for overall pressure drop show that

- Both local fluctuations near the pressure tap in the bed and fluctuations at the bed surface are present. These two types of dynamics can be separated by a two-dimensional decomposition, especially if the PC1 crossing plane is shifted toward the larger orbits in the trajectory.

- The maximum value of PC1 at the "J" crossing of MAJFM orbits is related to the void volume.

The phase-space trajectories for differential pressure measurements show that

- Only local voids and emulsion phase fluctuations appear to be represented.

- The orbits begin at high values and loop to lower values of PC1 as the void passes into the region between the taps. The trajectory reverts to the higher levels as the void passes out of the region between the taps.

Fluidization variations versus superficial velocity can be characterized by a two-dimensional decomposition and observation of the relative proportions of MNM, MAJFM, and MAJIJFM orbits. Near minimum fluidization, the fraction of time spent in MNM orbits is highest. At high superficial velocity, MAJIJFM orbits become more dominant.

Future Work

The methods we present in this paper are formative. They appear to be useful for characterizing trajectories in terms of physically significant properties of fluidized beds. Further application to other types of data will likely refine the methods and the interpretation of decomposition products.

LITERATURE CITED

1. Daw, C.S. and J.S. Halow, "Characterization of Voidage and Pressure Signals from Fluidized Beds Using Deterministic Chaos Theory," 1991 International Conference on Fluidized Bed Combustion, Vol. 2, pp. 777-786, April 21-24 (1991).

2. Daw, C.S. and J.S. Halow, "Modelling Deterministic Chaos in Fluidized Beds," AIChE Symposium Series 289, Vol. 88, pp. 61-69 (1992).

3. Daw, C.S. and J.S. Halow, "Evaluation and Control of Fluidization Quality Through Chaotic Time Series Analysis of Pressure-Drop Measurements," AIChE Symposium Series 296, Vol. 89, pp. 103-113 (1993).

4. Fuller, T.A., T.J. Flynn, C.S. Daw, and J.S. Halow, "Interpretation of Pilot-Scale, Fluidized Bed Behavior Using Chaotic Time Series Analysis," 12th International Conference on Fluidized Bed Combustion, Vol. 1, pp. 126-140 (1993).

5. Skrzycke, D.P., K. Nguyen, and C.S. Daw, "Characterization of the Fluidization Behavior of Different Solids Types Based on Chaotic Time Series Analysis of Pressure Signals," 12th International Conference on Fluidized Bed Combustion, Vol. 1, pp. 155-166 (1993).

6. Schouten, J.C. and C.M. van den Bleek, "Chaotic Behavior in a Hydrodynamic Model of a Fluidized Bed Reactor," 1991 International Conference on Fluidized Bed Combustion, Vol. 1, pp. 459-466 (1991).

7. Boulliard, J.X. and R.W. Lyczkowski, "Chaotic Nature of Gas/Solids Fluidization," AIChE National Meeting, Miami (1992).

8. Van der Stappen, M.L.M., J.C. Schouten, and C.M. van den Bleek, "Deterministic Chaos Analysis of the Dynamical Behavior of Slugging and Bubbling Fluidized Beds," 12th International Conference on Fluidized Bed Combustion, Vol. 1, pp. 126-140, May 9-13 (1993).

9. Broomhead, D.S. and D.P. King, *Physica 20D,* p. 271 (1986).

10. Jansen, B.H. and H.N. Nyberg, *Chaos*, Vol. 3, No. 2, pp. 143-152 (1993).

11. Halow, J.S., G.E. Fasching, P. Nicoletti, and J.L. Spenik, *Chem. Eng. Sci.*, Vol. 48, No. 4, pp. 643-659 (1993).

Table 1. Characteristics of Voids in the 0.604 m/s Experiment

Time (s)	Diameter (m)	Length (m)	Rise in Velocity (m/s)	Spacing Between Voids (m)
0.27	0.132	0.33	0.95	---
1.34	0.130	0.49	1.19	1.02
2.40	0.133	0.38	0.95	1.27
3.42	0.132	0.33	0.79	0.98
4.46	0.128	0.21	0.68	0.84
5.54	0.129	0.37	0.95	0.72
6.62	0.132	0.32	0.79	1.04
7.47	0.130	0.44	1.19	0.66
8.43	0.127	0.36	1.06	1.13
8.54	0.990	0.107	1.19	0.13
9.44	0.120	0.26	0.68	1.11
10.59	0.135	0.35	0.79	0.78
11.78	0.136	0.44	0.95	0.93
12.88	0.129	0.162	0.79	1.05
13.95	0.130	0.36	0.95	0.85
15.07	0.138	0.35	0.79	1.08
16.26	0.133	0.36	0.79	0.94
17.31	0.126	0.134	0.95	0.84
18.56	0.139	0.35	0.79	1.19
20.03	0.137	0.47	0.64	1.16

Table 2. Decomposition Parameters for Void Fraction and Overall Pressure Drop

Experiment	Orbit	Time in Orbit (%)	Average Period (s)	Frequency (1/s)
700 μm acrylic powder fluidized at 3.18 x Umf				
Void Fraction	MNM	19.8	0.27	0.75
	MAJFM	46.7	0.88	0.55
	MAJIJFM	29.5	0.88	0.35
Pressure Drop	MNM	50.9	0.20	2.55
	MAJFM	48.6	0.54	0.90

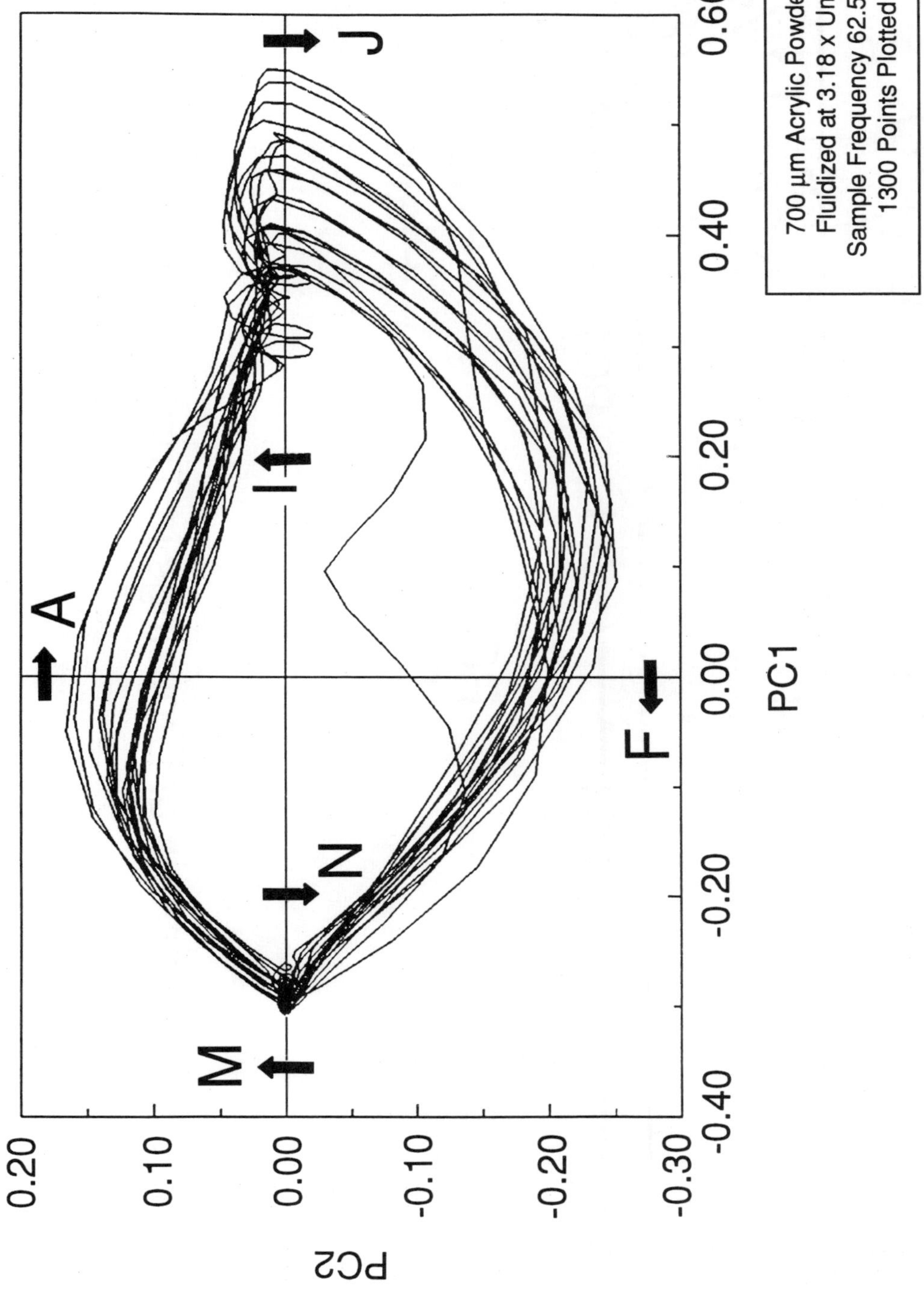

Figure 1. Phase Space Trajectory for Void Fraction Measurements

M94000505*

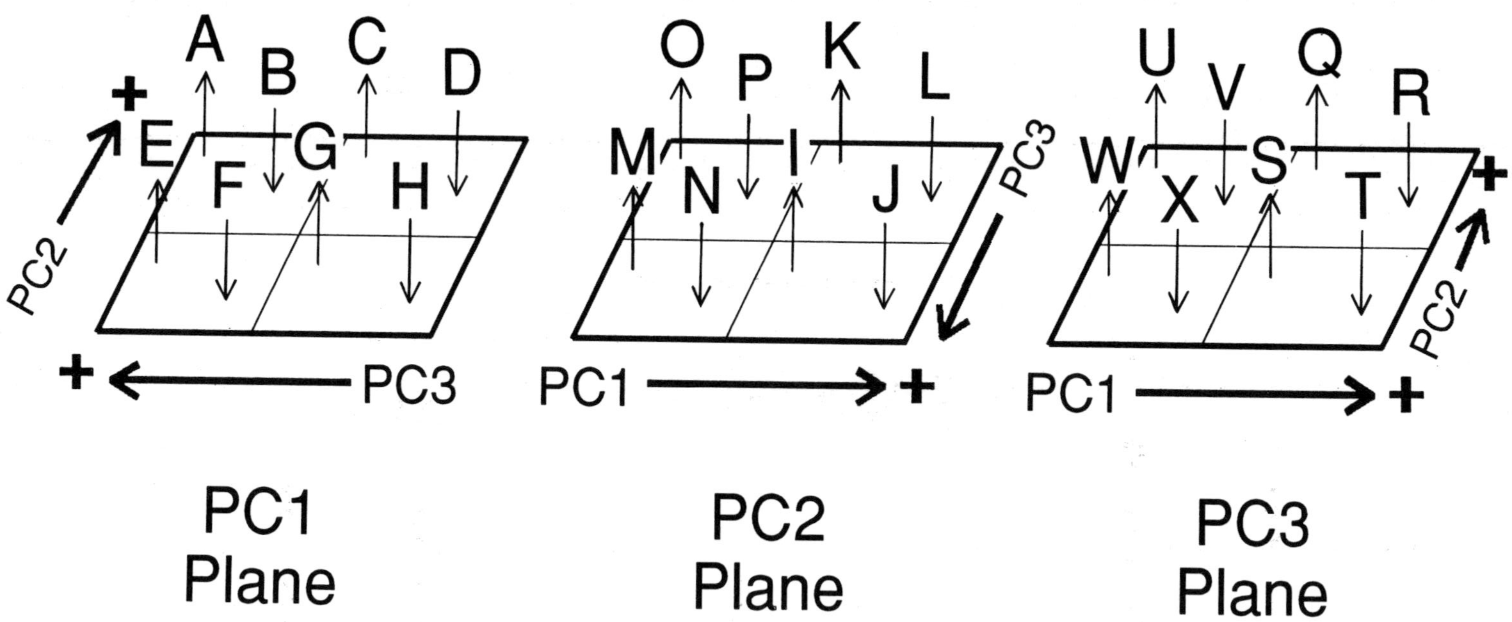

Figure 2. Definitions of Crossings in Three Dimensions

M94000507*

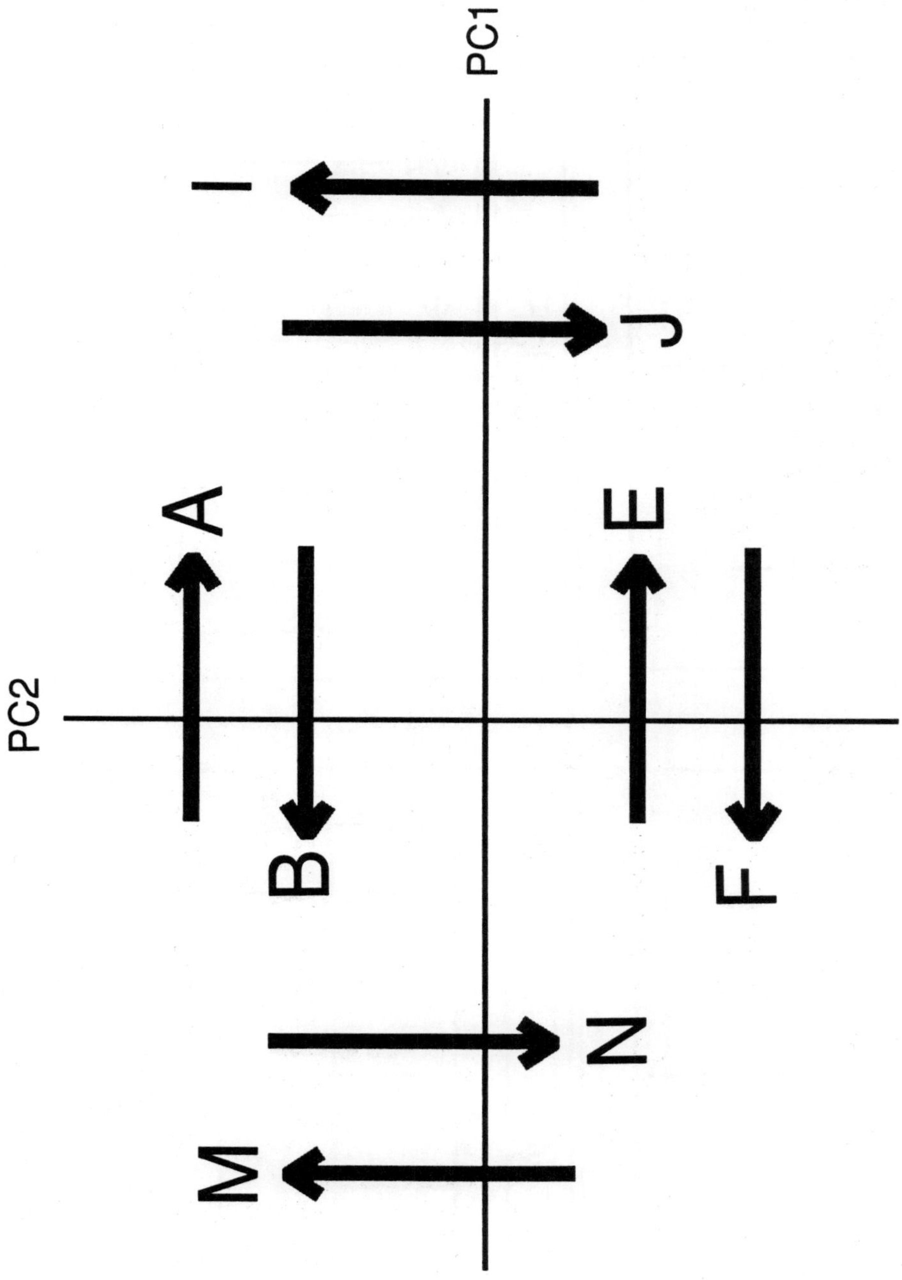

Figure 3. Definitions of Crossings in Two Dimensions

M94000508*

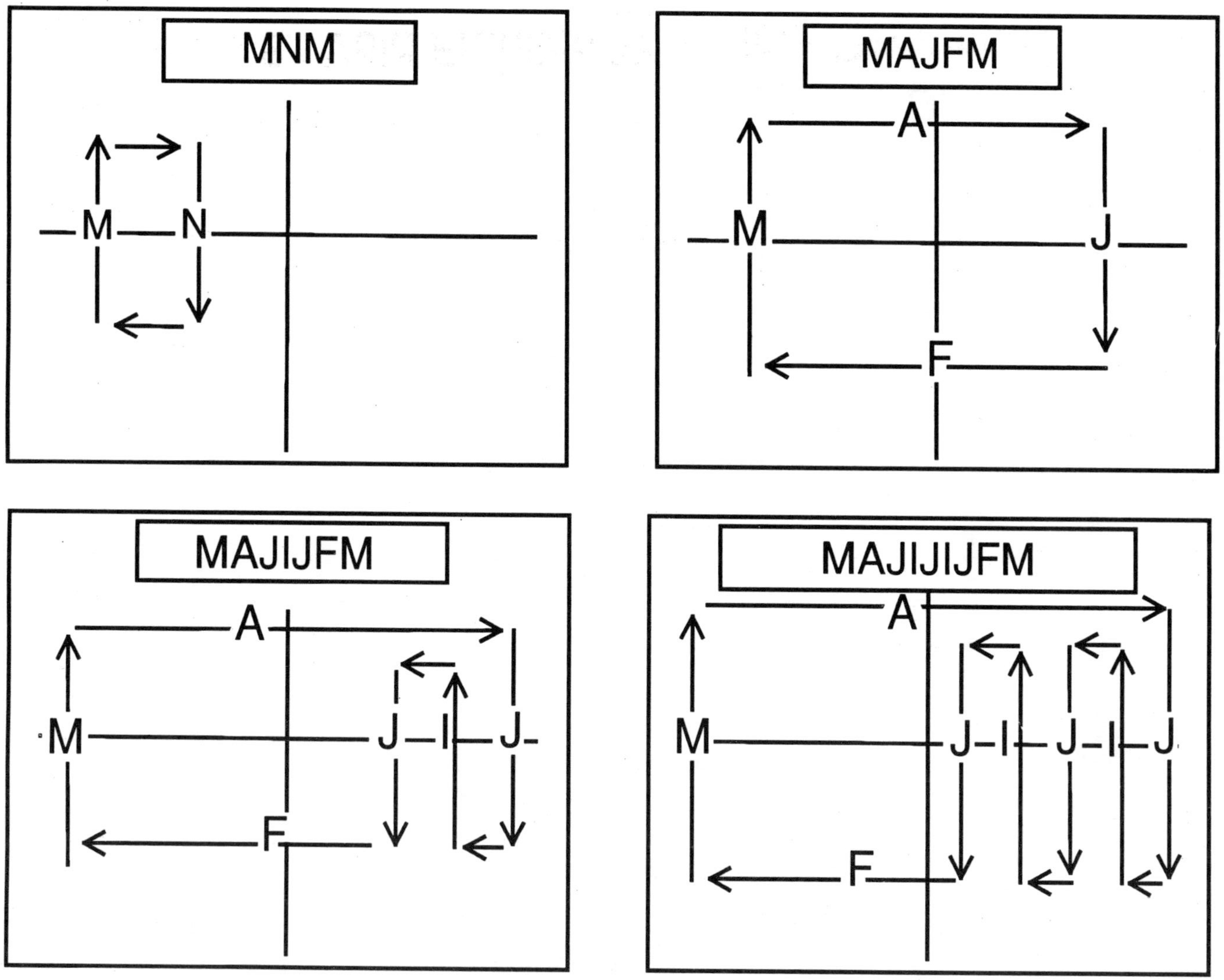

Figure 4. Illustrations of Common Crossing String Patterns

M94001517W

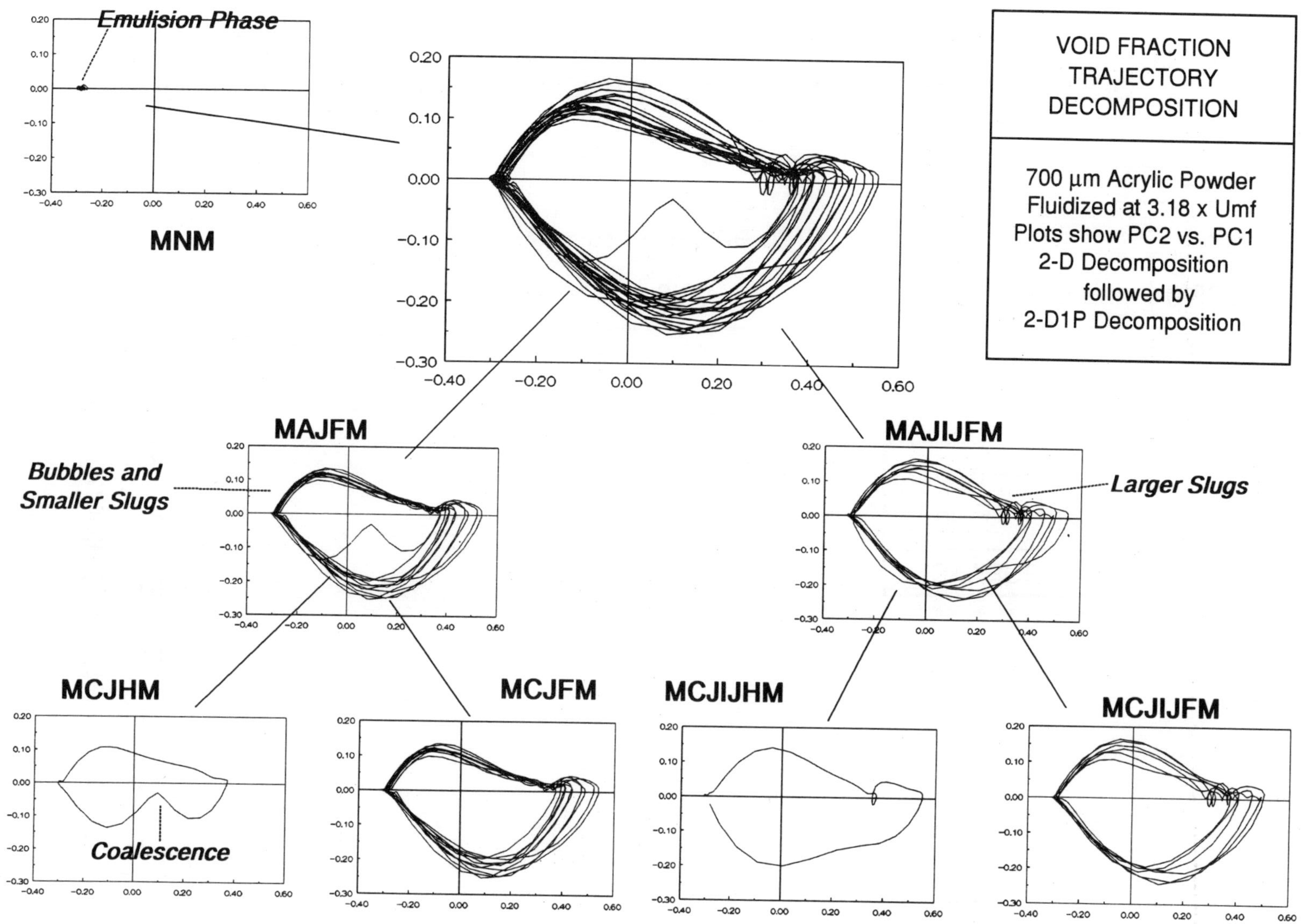

Figure 5. Void Fraction Trajectory Decomposition

M94000511*

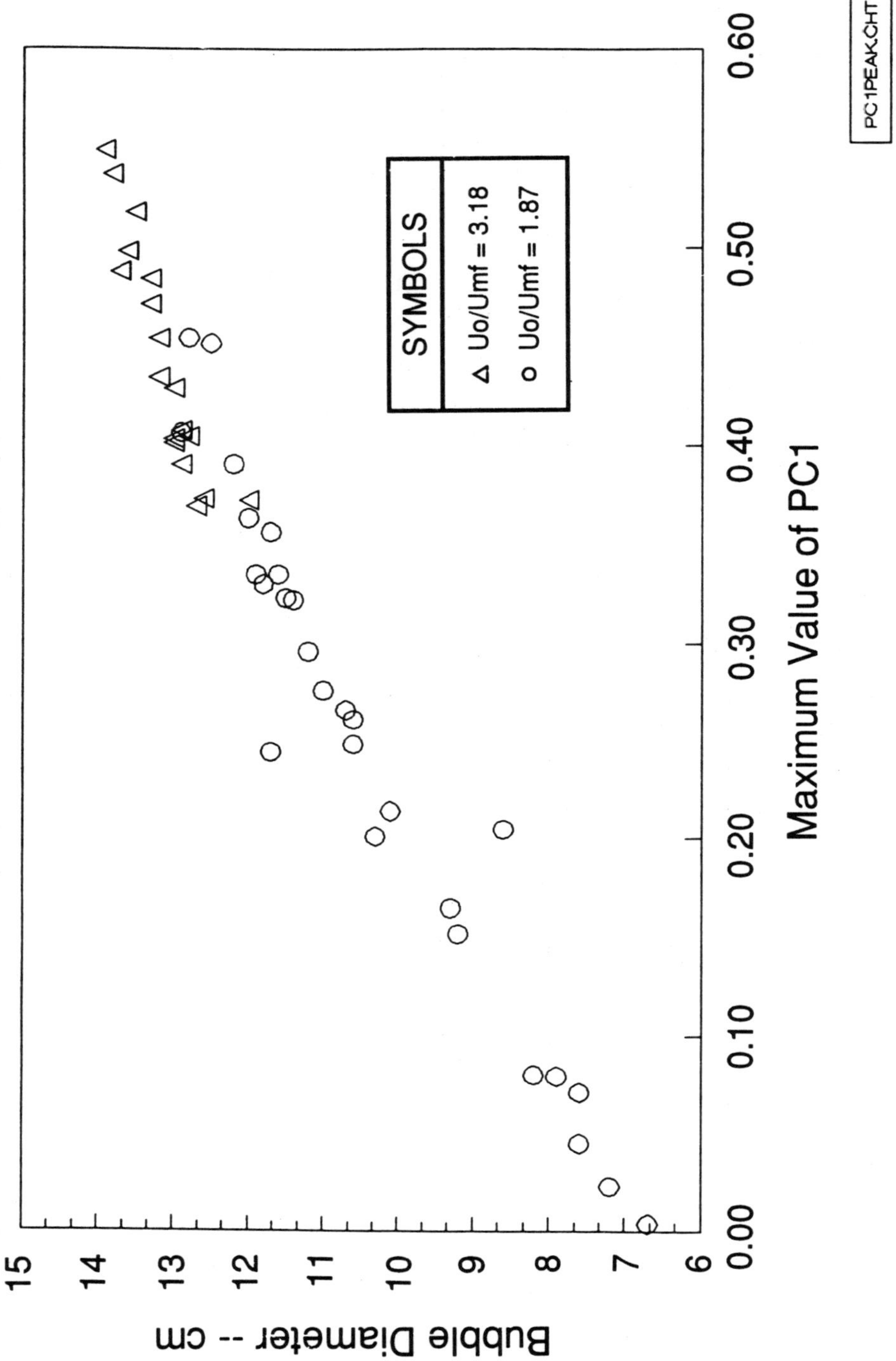

Figure 6. Relationship Between Bubble Diameter and PC1 Maxima in Void Fraction Measurements

M94000512*

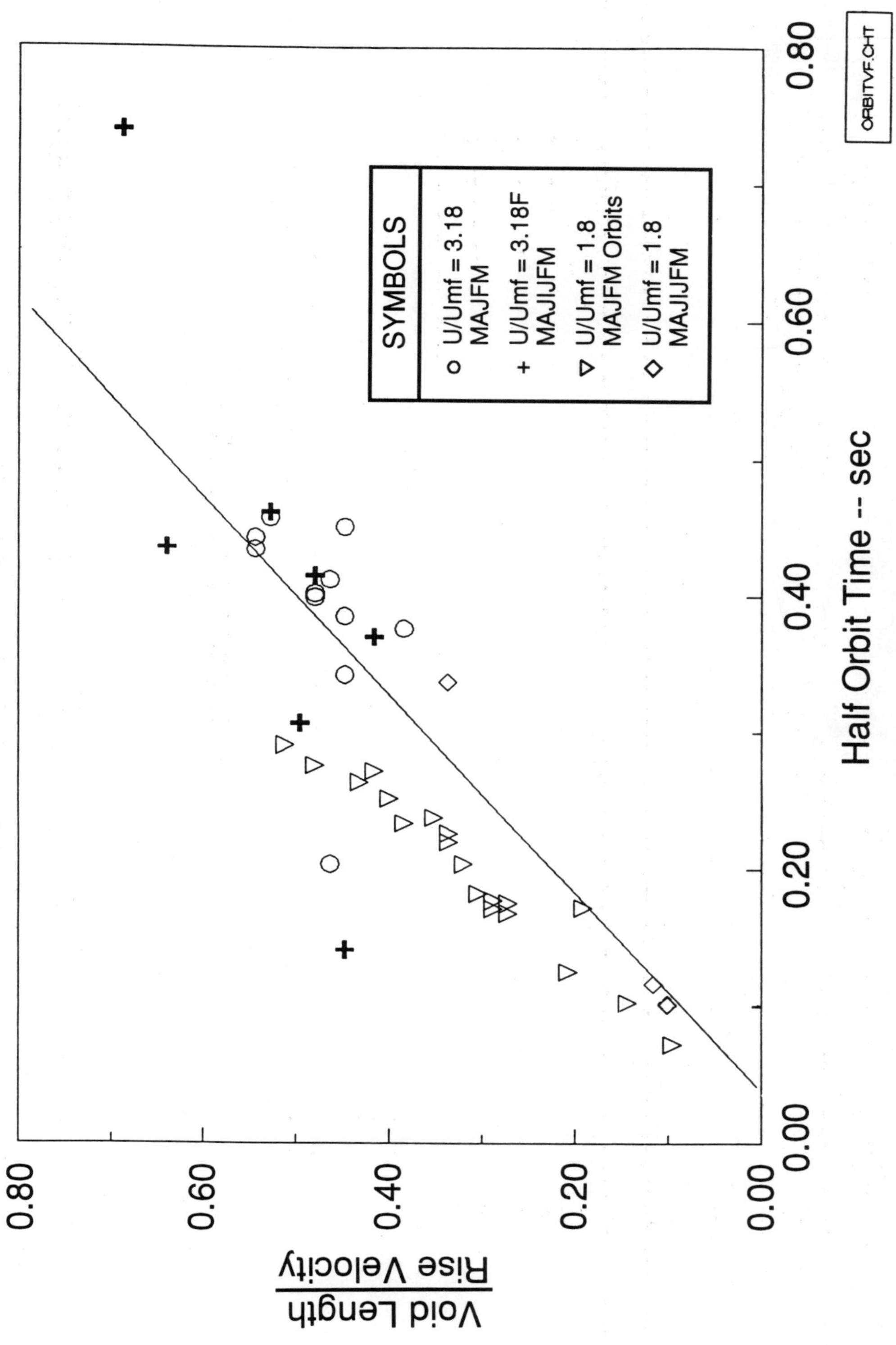

Figure 7. Relationship of Half Orbit Time and Void Properties in Void Fraction Measurements

M94000513*

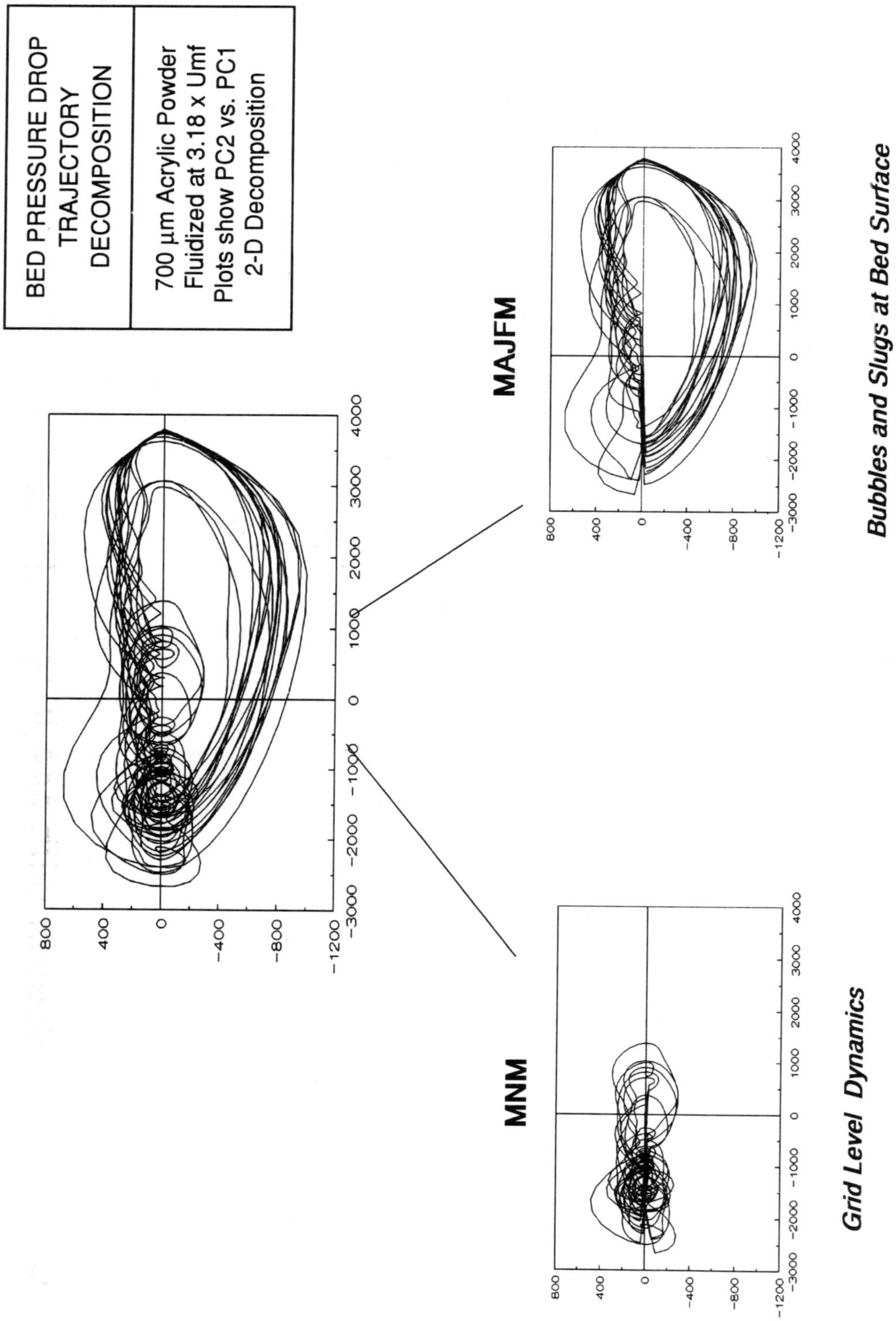

Figure 8. Bed Pressure Drop Trajectory Decomposition

M94000514*

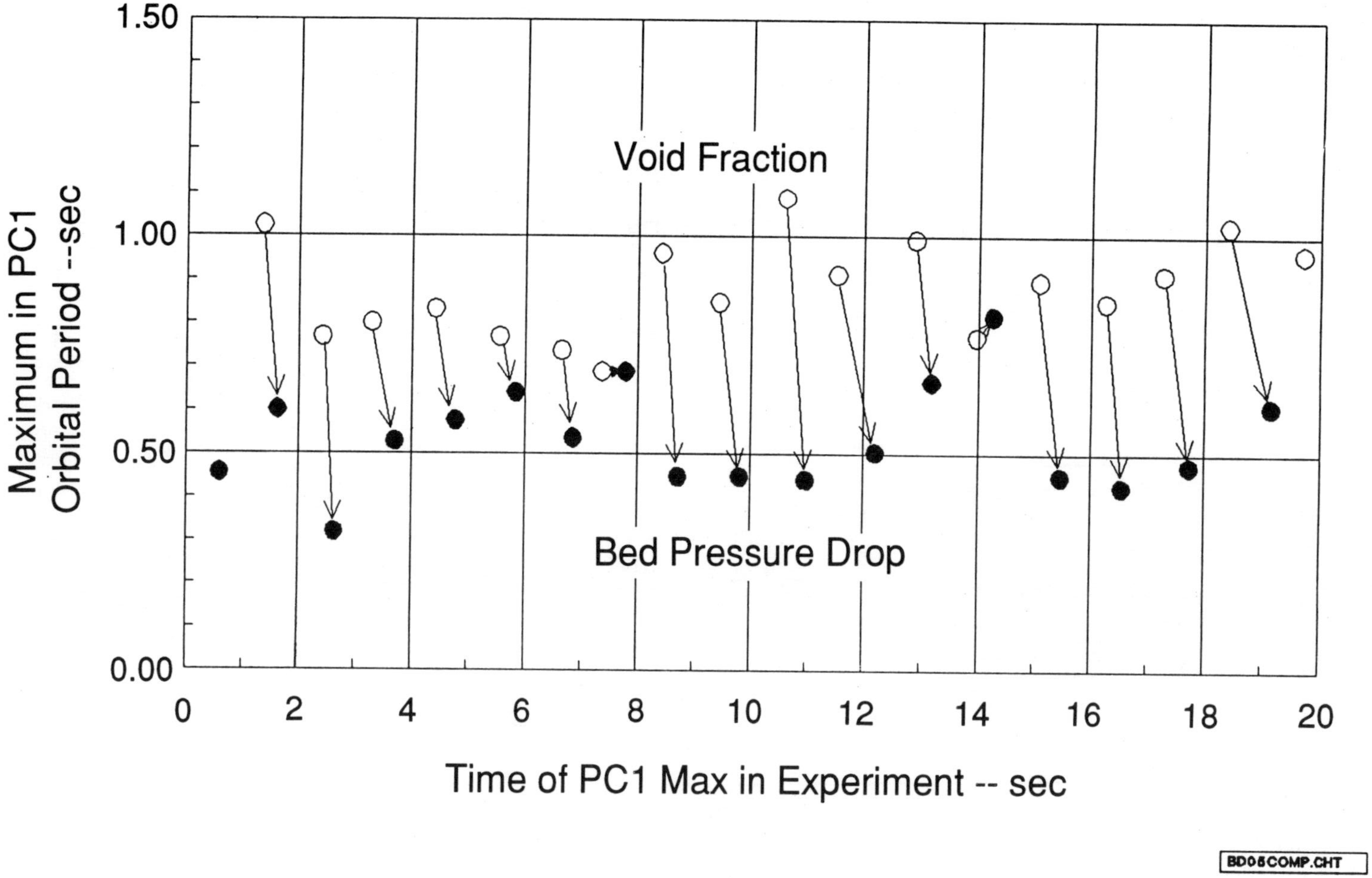

Figure 9. Comparison of Void Fraction and Bed Pressure Drop PC1 Maxima with MAJFM and MAJIJFM Orbits

M94000515*

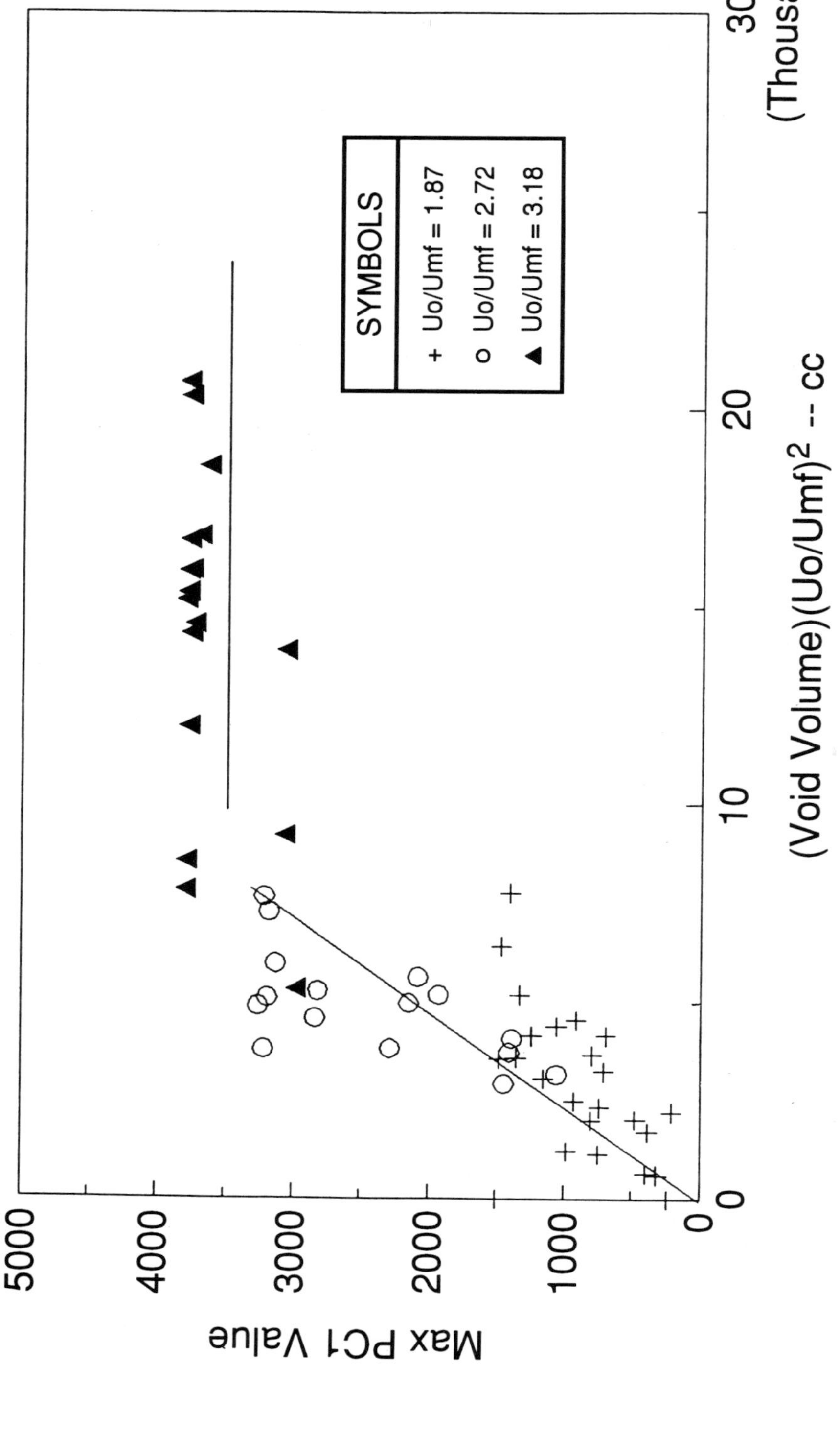

Figure 10. Relationship Between Bed Pressure Drop PC1 Maxima and Void Volume

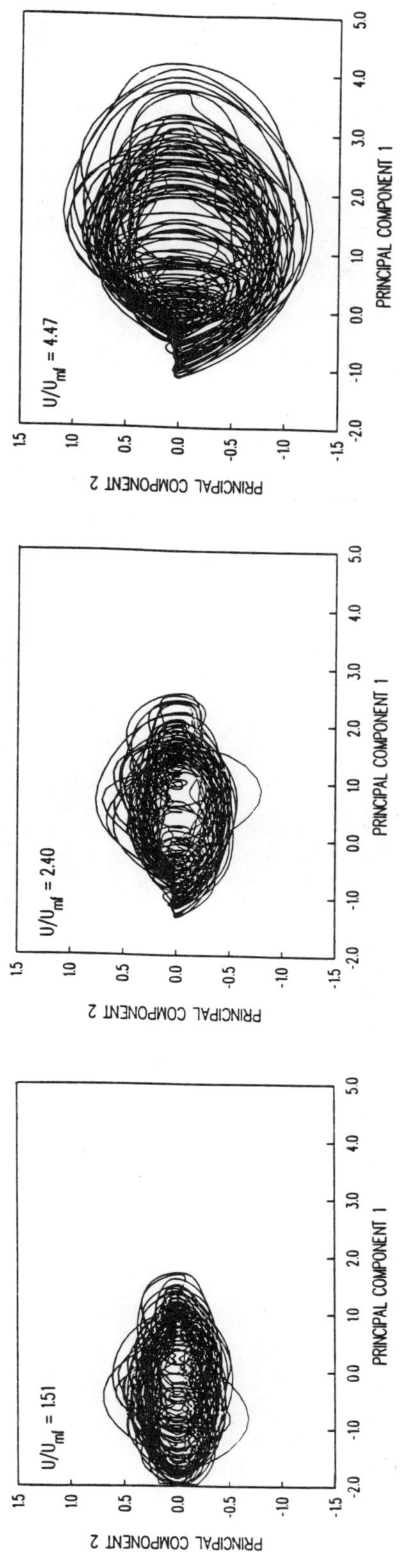

Figure 11. Phase Space Trajectories for Local Pressure Drop Measurements in Various Fluidization Regimes

M94000518*

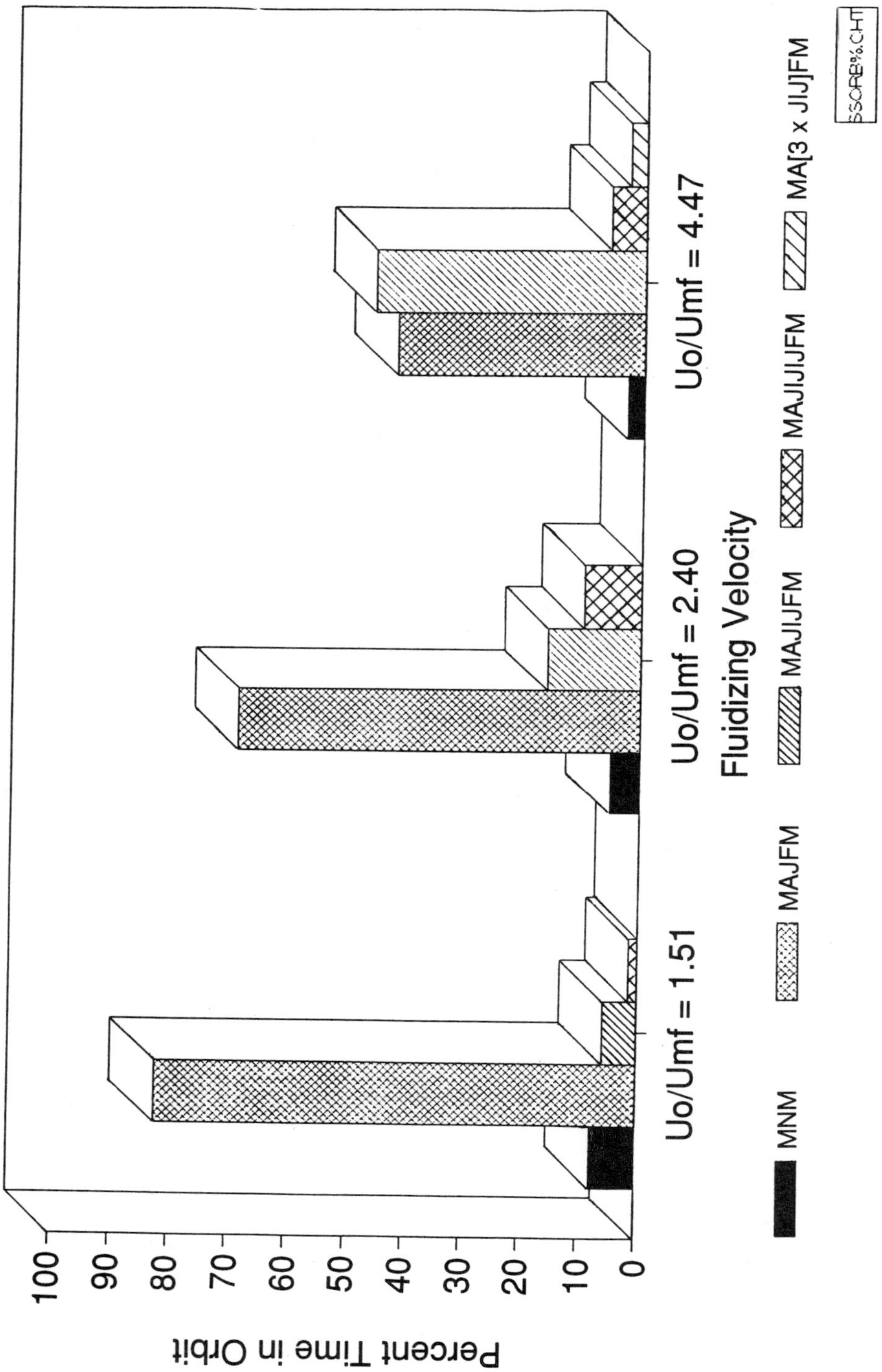

Figure 12. Changes in Percent of Time in Various Orbital Groups with Fluidizing Condition

M94000519*

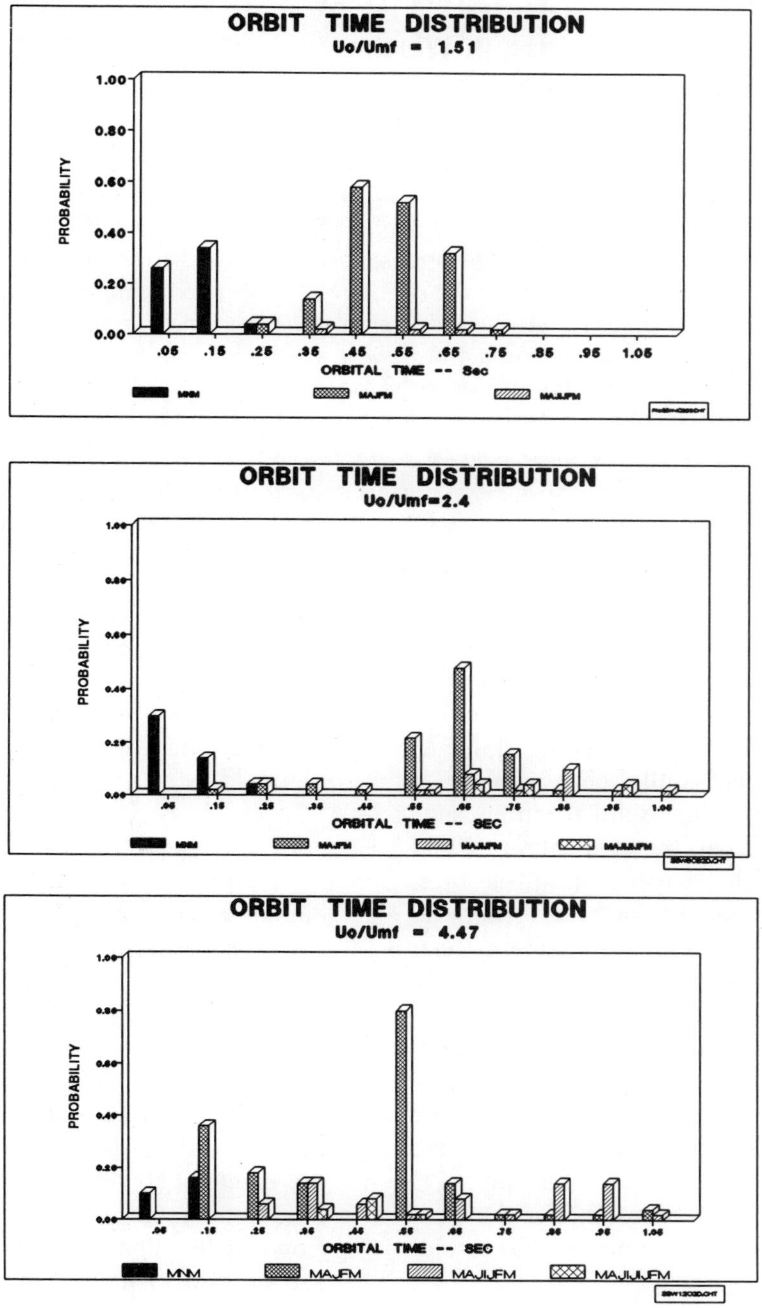

Figure 13. Distribution of Orbital Periods Under Various Fluidizing Conditions

M94000520*

A Multi-Stage Model for the Circulating Fluidized Bed Reactor

H. Schoenfelder, J. Werther, J. Hinderer and F. Keil
Technical University Hamburg-Harburg, Denickestr.15, D - 21071 Hamburg, Germany

The present work shows an approach to the development of a reaction model for the circulating fluidized bed reactor that is based on the consideration of the local flow structure. The sequence of different fluid dynamical zones that can be found in the riser reactor leads to its subdivision into four stages with different model formulations. Recent work has shown that the bottom zone of the CFB riser can successfully be described by modified bubble models. The upper part of the riser, the so-called recirculation zone, is separated from the denser bubble region by the splash zone. The exit region of the reactor yields a mixing condition. The recirculation zone is described by a one-dimensional two-phase model that is based on the assumption of a countercurrent flow situation for falling clusters (dense phase) and the upflowing suspension (lean phase). Mass exchange in the recirculation zone between the phases is characterized by a mass transfer coefficient k [m/s] and the volume specific interfacial mass transfer area a [m^2/m^3]. The resulting mass exchange coefficient $k \cdot a$ was determined by tracer gas backmixing experiments and model calculations. These experiments were conducted in a cold flow riser (height: 11m, dia.: 0.03 m) with CO_2 and argon as tracer gases. The particles used were quartz sand (d_p=80 μm) and catalyst (dp=100 μm). Values of $k \cdot a$ in the range of 1.5 to 2 [1/s] were obtained. Knowing $k \cdot a$, basic geometric data and the reaction kinetics, the four-stage model yields a complete description of the reactor behavior. As an example the model was used to predict the performance of a CFB reactor for the catalytic conversion of methanol to olefins.

During the last two decades circulating fluidized bed reactors have successfully been applied to numerous processes with gas-solid reactions such as the combustion of coal, the calcination of aluminum hydroxide or the roasting of sulfide ores. Contrary to this development, for heterogeneously catalyzed reactions, there is only a limited number of applications to be found, e.g. the fluid catalytic cracking of heavy oil (FCC). Circulating fluidized bed reactors offer two major advantages for catalytic processes:

- Spent catalyst is quickly removed from the reactor and can continuously be regenerated. This feature is very important for the FCC process and could also be useful for the conversion of methanol to gasoline (MTG) or olefins (MTO) [1].
- Circulating fluidized bed reactors offer the opportunity to use the solids recirculation line for adsorbing oxygen to the catalyst's surface. The selectivity of some catalytic reactions is believed to be improved by this way of feeding reactants. For example, the production of maleic anhydride from oxygen and butane could be an important application of this new technology [2].

Despite these advantages the application of circulating fluidized beds for catalytic processes still suffers from the problem of high catalyst attrition and the lack of comprehensive and predictive models [3].

When scientists started their research on circulating fluidized beds in the 1970s, they assumed the gas to be in plug flow while passing through the reactor [4]. As a result, reactor models were based on simple approaches. Contrary to the plug flow assumption following investigations have shown a high extent of axial dispersion within the riser, e.g. [5]. These findings lead to the development of improved dispersion models, some of which were able to account for the fact that the flow structure in the riser has axial nonuniformities. EDWARDS and AVIDAN [6] therefore introduced two different axial dispersion coefficients for the lower and upper part of the reactor. Local measurements in the upper part of the riser

Technical University Hamburg-Harburg
Denickestr. 15, D-21071 Hamburg, Germany

of a circulating fluidized bed [7] indicated that it is an internal recirculation of gas and solids that effects the axial dispersion.

At present, much more about the complex flow structure in circulating fluidized beds is known. Investigations of many authors have shown that there exists a bottom zone with a high solids concentration. In the upper part of the riser the solids concentrations are much lower than in the bottom zone. Between these regions an exponential decay of the solids concentration has been observed. In the upper dilute flow region researchers found a core-annulus kind of flow structure that gives an explanation of the axial dispersion phenomena: In the annulus zone, near the wall, downfalling clusters or strands give rise to intense backmixing of gas and solids. Meanwhile, extensive experimental investigations as well as models have been published in the area of CFB fluid mechanics and gas mixing (e.g. [8], [9], [10], [11]).

Though a lot of information is available about the flow structure in circulating fluidized beds, there are still only few reactor models to be found that account for these findings. KAGAWA et al. [12] described the catalytic decomposition of ozone by a model that was based on the assumption of a stagnant annulus zone with a high solids content. Another model of a circulating fluidized bed reactor for the conversion of butane to maleic anhydride was presented by PUGSLEY et al. [13]. They presumed the annulus zone to have a solids concentration equal to the state of minimum fluidization. Though the solids in the annulus are assumed to move downwards with the single particle's terminal velocity the gas in this zone is regarded as stagnant. Gas exchange between the core and the annulus is described by a coefficient of mass transfer.

The modelling efforts presented here were focused on the development of a description of the riser that accounts for the sequence of different fluid mechanical zones in the axial direction and the backmixing phenomena in the upper part of the riser. As a first step, a one-dimensional modelling approach is proposed in the present paper which should give a description of the overall reactor behavior.

THE CFB REACTOR MODEL

Contrary to most hitherto existing models which either distinguish between bottom/dense and upper/dilute zones (e. g. [14]) or concentrate on the description of the core/annulus flow structure (e. g. [15]), the present approach is supposed to give a comprehensive description of the overall reactor behavior. The riser is therefor subdivided into a sequence of four axial zones. Within these stages two separate phases are used to describe the flow structure.

Axial staging

In order to account for the axial nonuniformities in the riser, the circulating fluidized bed reactor is subdivided into four stages with different model formulations. Figure 1 shows the arrangement and linkage of these stages.

The first stage comprehends the bottom zone of the reactor where usually the highest solids concentrations are found. LECKNER and coworkers [14] and WERTHER and WEIN [17] have shown that this zone can be described by bubble models that were basically derived for low velocity fluidized beds.

Above the bottom zone the so-called splash zone is defined for the region where bubbles and falling strands dash against each other. Due to these collisions a high extent of mixing is assumed in this part of the riser. Because the splash zone covers only a small fraction of the reactor volume its importance for the reactor is not due to the chemical reactions taking place there, but to the mixing condition that it yields for the linkage of the bottom zone to the upper part of the riser, the so-called recirculation zone.

This recirculation zone is characterized by the coexistence of a lean upflowing suspension and downfalling strands or clusters that effect a recirculation of gas and solids within the riser. The axial profile of the solids concentration in the recirculation zone can be predicted e.g. by the equations of KUNII and LEVENSPIEL [8] or WERDERMANN [19]. For the description of gas-solid contacting in the recirculation zone a new one-dimensional model has been developed that is presented in this paper.

In the exit region it can be assumed that the flow of gas and solids is characterized by a high extent of mixing. Since riser reactors usually have slim geometries, the exit region provides a rather small amount of reaction volume. Thus, the exit region is simply assumed to yield a mixing condition for the recirculation zone and chemical reactions are neglected.

The sub-model for the recirculation zone

Usually most of the riser height is occupied by the recirculation zone. Thus, a significant fraction of the chemical conversion may be expected to take place in this part of the reactor, even though solids concentrations are lower than in the bottom zone. Modelling of the recirculation zone must account for an important feature of the flow structure that has been observed: The influence of falling strands on the reactor performance.

As can be seen from Figure 2, two phases are introduced to represent the clusters/strands and the upflowing suspension. The dense phase and the lean phase are assumed to be in a counter-current flow relationship. The fraction f_d of the cross sectional area is occupied by the dense phase. It is described by concentrations of gas components C_{id}, the solids volume concentration c_{vd} and the velocities of gas and solids u_d and v_d, respectively. The lean phase is characterized by a similar set of variables: C_{il}, c_{vl}, u_l and v_l. Mass exchange between the phases is described by an area based mass transfer coefficient k [m/s] and the specific interfacial area of the phases a [m²/m³] that is based on the reactor volume.

Determining the fluid mechanical variables as a function of a given set of operating conditions is based on the assumption that the solids circulation rate G_s and the axial profile of solids concentration $\bar{c}_v$ are given. The solids concentration of the dense phase c_{vd} and the dense phase solids velocity v_d can be obtained by direct measurements, e.g. with optical probes [24]. For the calculation of the remaining five variables f_d, c_{vl}, u_l, v_l, u_d a system of five equations can be solved analytically.

In the lean phase, the relative velocity between particles and gas is assumed to be equal to the equivalent terminal settling velocity for a single particle in the gas,

$$u_l - v_l = \sqrt{\frac{4 \cdot (\rho_s - \rho_g) \cdot g \cdot d_p}{3 \cdot c_w(\mathrm{Re}) \cdot \rho_g}} \tag{1}$$

where the drag coefficient c_w is calculated according to the equation of BRAUER and KASKAS [20].

The relationship of gas and particle velocities in the dense phase is assumed to obey the equation of RICHARDSON and ZAKI [21].

$$u_d - v_d = \frac{u_{mf}}{1 - c_{v,mf}} \cdot \left(\frac{1 - c_{vd}}{1 - c_{v,mf}} \right)^{n-1}, \quad n = \mathrm{f}(\mathbf{Re}_t) \tag{2}$$

The definition of the cross-sectional average solids volume concentration $\bar{c}_v$, Equation (3), and the balances of gas, Equation (4), and solids flow, Equation (5), yield three additional equations.

$$\bar{c}_v = f_d \cdot c_{vd} + (1 - f_d) \cdot c_{vl} \tag{3}$$

$$u = u_l \cdot (1 - f_d) \cdot (1 - c_{vl}) + u_d \cdot f_d \cdot (1 - c_{vd}) \tag{4}$$

$$G_s = \rho_s \cdot v_l \cdot c_{vl} \cdot (1 - f_d) + \rho_s \cdot v_d \cdot c_{vd} \cdot f_d \tag{5}$$

Once the fluid mechanical variables are calculated, differential balances for the gas components can be evaluated. The resulting balance equations which can be derived from Figure 3 are based on the following assumptions:

- axial or radial dispersion within the phases is negligible
- riser diameter is constant
- gas in the dense phase has constant density
- velocities of gas and solids of the dense phase are constant in time and invariant with height
- axial profile of superficial gas velocity is constant in time
- mass transport of gas components by adsorption to the solids surface is negligible

In case of non-steady state conditions the balances for n gas components yield a system of $2 \cdot n$ partial differential equations:

dense phase:

$$\frac{\delta C_{id}}{\delta t} = -u_d \cdot \frac{\delta C_{id}}{\delta h} - \frac{u_d}{f_d} \cdot C_{id} \cdot \frac{\delta f_d}{\delta h} + \frac{c_{vd}}{(1-c_{vd})} \cdot \rho_s \cdot R_{id} - k \cdot a \cdot \frac{1}{f_d \cdot (1-c_{vd})} \cdot (C_{id} - C_{il}) + \frac{\dot{V}_{conv} \cdot C_{conv}}{f_d \cdot (1-c_{vd}) \cdot A_t} \quad (6)$$

lean phase:

$$\frac{\delta C_{il}}{\delta t} = \frac{c_{vl}}{1-c_{vl}} \cdot \rho_s \cdot R_{il} + C_{il} \cdot \frac{u_l}{(1-c_{vl})} \cdot \frac{\delta c_{vl}}{\delta h} + k \cdot a \cdot \frac{(C_{id} - C_{il})}{(1-f_d) \cdot (1-c_{vl})} - u_l \cdot \frac{\delta C_{il}}{\delta h} - \frac{C_{il}}{1-f_d} \cdot \left(\frac{\delta u_l}{\delta h} \cdot (1-f_d) - \frac{\delta f_d}{\delta h} \cdot u_l\right) - \frac{\dot{V}_{conv} \cdot C_{conv}}{(1-f_d) \cdot (1-c_{vl}) \cdot A_t} \quad (7)$$

where the convective exchange between the phases can be calculated from:

$$\dot{V}_{conv} = u_d \cdot (1 - c_{vd}) \cdot A_t \cdot \left(\frac{\delta f_d}{\delta h} + \frac{\delta P}{\delta h} \cdot \frac{f_d}{P}\right) - \frac{\Re \cdot T}{P} \cdot f_d \cdot A_t \cdot c_{vd} \cdot \rho_s \cdot \sum R_{id} \quad (8)$$

$$C_{conv} = \begin{Bmatrix} C_{il} & (\dot{V}_{conv} > 0) \\ C_{id} & (\dot{V}_{conv} < 0) \end{Bmatrix} \quad (9)$$

CFB reactor model equations and boundary conditions

The reactor behavior is described by systems of differential equations that are derived from mass balances for each component. Since different fluid mechanical approaches are required for the axial zones of the riser, different systems of differential equations are used that have to be linked by defining appropriate boundary conditions.

For the bottom zone of the reactor a system of ordinary differential equations describes the steady state gas phase composition as a function of height.

bubble phase:

$$\frac{dC_{ib}}{dh} = \frac{1}{u - u_s \cdot (1 - \varepsilon_b)} \cdot \left(-k_{ig} \cdot a \cdot (C_{ib} - C_{is}) + K_Q \cdot C_{is} - C_{ib} \cdot \frac{du}{dh} - u_s \cdot C_{ib} \cdot \frac{d\varepsilon_b}{dh}\right) \quad (10)$$

suspension phase:

$$\frac{dC_{is}}{dh} = \frac{1}{u_s \cdot (1 - \varepsilon_b)} \cdot \left(k_{ig} \cdot a \cdot (C_{ib} - C_{is}) + K_Q \cdot C_{is} + u_s \cdot C_{is} \cdot \frac{d\varepsilon_b}{dh} + \rho_s \cdot (1 - \varepsilon_b) \cdot c_{vs} \cdot R_i\right) \quad (11)$$

where the convective exchange between bubbles and suspension, K_Q, can be calculated by Equation (12).

$$K_Q = u_s \cdot \frac{d\varepsilon_b}{dh} - u_s \cdot (1-\varepsilon_b) \cdot \frac{dP}{dh} \cdot \frac{1}{P} + \Re \cdot T \cdot \frac{1}{P} \cdot \rho_s \cdot (1-\varepsilon_b) \cdot c_{vs} \cdot \sum R_i + \Re \cdot T \cdot \frac{1}{P} \cdot \sum \left[k_{ig} \cdot a \cdot (C_{ib} - C_{is}) \right] \quad (12)$$

The mass transfer coefficient k_{ig} is calculated according to SIT and GRACE [18] and the axial profile of fluid mechanical variables, such as bubble diameter and bubble gas hold-up, can be determined by the correlations of WERTHER and WEIN [17].

The Equations (10) and (11) yield a system of ordinary differential equations with an initial condition at the gas distributor:

$$h = 0 \Rightarrow C_{is} = C_{ib} = C_{if} \quad (13)$$

The sub-model of the recirculation zone (Equations (6) and (7)) is more difficult to describe than the bottom zone model, because the counter-current flow approach requires two-point boundary conditions. These boundary conditions are provided by the splash-zone, Equation (14), and the exit region, Equation (15):

$$h = H_B + H_{sz} \Rightarrow C_{il} = C_{i,sz} \quad (14)$$

$$h = H \Rightarrow C_{id} = C_{il} \quad (15)$$

The concentration $C_{i,sz}$ of component i in the splash zone is determined by Equation (16),

$$\frac{dC_{i,sz}}{dt} = \frac{1}{H_{sz} \cdot (1-c_{V,sz})} \cdot \big(H_{sz} \cdot c_{V,sz} \cdot \rho_s \cdot R_i + u_s \cdot (1-\varepsilon_b) \cdot C_{is} + (u - u_s \cdot (1-\varepsilon_b)) \cdot C_{ib} + f_d \cdot (1-c_{Vd}) \cdot u_d \cdot C_{id} - (1-f_d) \cdot (1-c_{Vl}) \cdot u_l \cdot C_{i,sz} \big) \quad (16)$$

where all height depending variables originating from the bottom and the recirculation zone must be set to the values at $h=H_B$ and H_B+H_{SZ}, respectively.

Non-steady state calculations for the recirculation zone and the splash zone are performed with the initial condition $t=0$: $C_i=C_{ib}$ $(h=H_B)$ for both phases and the splash zone. The system of partial differential equations was solved either with the method of lines [22] or the method of characteristics [23].

In case of steady state conditions, the balances form a system of ordinary differential equations with two-point boundary conditions provided by the splash zone and the exit region. The stability and convergence of usual numerical algorithms were in some cases unsatisfactory. Instead, the alternative of integrating the non-steady state equations until steady state conditions were reached was used successfully.

EXPERIMENTAL

A solution of the model equations depends on the knowledge of fluid mechanical variables, such as G_s and the axial profile of $\bar{c}_V$, as a function of operating conditions. Whereas for prediction of most fluid mechanical variables approaches have been published, there is no means available yet from literature to obtain values for $k \cdot a$.

It was therefore decided to determine $k \cdot a$ from separate gas mixing experiments in a cold flow model. The riser of the apparatus shown in Figure 4 is 11 m in height and has an inner diameter of 30 mm. The slow bed in the solids return line has an inner diameter of 100 mm and was equipped with a cone valve for control of the solids mass flux which was measured by the volume method. Eleven pressure taps along the riser were used to determine the axial profile of the apparent solids concentration $\bar{c}_V$.

The solids used were either sand particles (d_p: 80 µm, ρ_s: 2600 kg/m³, u_{mf}: 10 mm/s) or catalyst (d_p: 100 µm, ρ_s: 1500 kg/m³, u_{mf}: 8 mm/s). The catalyst is a zeolite of the HZSM-5 type that has been provided by the Vereinigte Aluminium Werke. The solids were fluidized with ambient air, the moisture of which was controlled in order to reduce electrical charges that could have disturbed measurements with optical probes.

Direct measurements for determination of solids concentration and solids velocity have been carried out with optical probes. They are of the single fibre reflection type. The calibration procedure as well as the signal evaluation are described by RENSNER and WERTHER [24]. The probe can measure solids concentrations and velocities simultaneously by using a pair of fibres, the concentration signals of which are cross-correlated in order to obtain velocities.

The description of gas mixing by the present model for the recirculation zone depends on the knowledge of the parameter $k{\cdot}a$. For determination of this variable, tracer gas backmixing experiments were conducted. A part of the riser was therefore equipped with a number of additional openings that are shown in Figure 5. Tracer gas was injected radially inwards. The axial profile of tracer gas concentration was measured with gas sampling probes.

Probes have been used which had a tip made of a sintered metal cylinder in order to prevent solids from entering the suction line. The tracer gas used was either CO_2 or argon. As far as the transport of gas is concerned adsorption phenomena are assumed to be negligible, because the temperatures which are typical for catalytic reactions usually reduce the catalysts' adsorption capacities. Since the zeolite catalyst shows high adsorption capacity for CO_2, argon was used as a tracer for the experiments carried out with catalyst particles. A Leybold Heraeus BINOS type infrared analyser was used for the CO_2 and argon was detected by a Leybold Heraeus mass spectrometer.

Since both analysers require only very low sampling flow rates, disturbances to the flow structure within the riser were limited to those caused by the presence of the probe. The respective probes were inserted at the time the measurements started, in order to reduce shielding by probes that were not in use. After a probe was moved to the desired position, gas was sucked off by a vacuum pump and fed to the IR- or the mass spectrometer, respectively. The signals provided by these analysers were recorded digitally on a personal computer for at least two minutes. Each measuring point was obtained as the average value of this measuring time.

RESULTS AND DISCUSSION

Solids concentration and velocity

Local measurements with the optical probe showed a wide distribution of solids velocities in the recirculation zone of the riser. The average value of the negative (downwards) velocities was about -0.75 m/s. The corresponding solids concentration in the strands was found to be very difficult to obtain, because the time averaged local solids concentration measured is always a combination of both, the dense and the lean phase. Assuming the peaks of the probe's signal records to represent the dense phase solids concentration, a value of c_{vd} = 25% has been obtained as a reasonable estimation.

Gas mixing experiments

Evaluation of the backmixing experiments was found to suffer from a general difficulty concerning the interpretation of the measured concentrations. The model calculations yield both, the concentration of tracer in the dense phase C_d and in the lean phase C_l. The sampling probe, on the other hand, always sucks gas from both, the dense phase and the lean phase. Thus, the concentrations measured always represent a mixture of both phases. Evaluating the resulting signal requires a method to determine the fraction of gas in the sample flow which originates from the dense phase and the lean phase, respectively. Due to the narrow riser diameter the available optical probe can not be applied to direct measurements of the radial solids concentration. It is therefore assumed that the radial distribution of solids obeys the equation of ZHANG et al.[25].

$$(1-c_v)=(1-\bar{c}_v)^{(0.191+r^{2.5}+3 \cdot r^{11})} \tag{17}$$

For a given dimensionless radial position r, the local fraction of the dense phase f_{dr} amounts to

$$f_{dr}(r)=\frac{c_v(r)-c_{vl}}{c_{vd}-c_{vl}} \tag{18}$$

The sampling gas tracer concentration provided by the probe, C_{pro}, is assumed to equal the integral of the local concentrations along the length L_{pro} of the probe's sintered metal cylinder.

$$C_{pro} = \frac{1}{L_{pro}/R} \int_{\frac{(R-L_{pro})}{R}}^{1} \left[f_{dr}(r) \cdot C_d + (1-f_{dr}(r)) \cdot C_l \right] \cdot dr \tag{19}$$

Since all the injected tracer gas leaves the reactor with the main flow, the net flux of tracer gas below the injection level is zero,

$$f_d \cdot u_d \cdot (1-c_{vd}) \cdot C_d + (1-f_d) \cdot u_l \cdot (1-c_{vl}) \cdot C_l = 0 \tag{20}$$

Equations (19) and (20) yield a complete means to calculate C_d and C_l from a given value of C_{pro} and vice versa.

Two examples of the backmixing experiments are shown in Figures 6 and 7. Using the concentration of the top sampling probe as an initial condition, model calculations were performed for different values of the parameter $k{\cdot}a$. The curves in Figures 6 and 7 result from a best fit calculation.

Fitting of the unknown parameter $k{\cdot}a$ was carried out with the SIMPLEX-method [27] and yielded values of 1.5 1/s for the catalyst/argon system and 1.8 1/s for sand/CO_2. Although the values obtained seem to be almost equal, the conclusion that $k{\cdot}a$ is a constant should not be drawn. Many open questions concerning the dependence of $k{\cdot}a$ on the operating conditions, the solids properties and the reactor geometry are yet to be answered. Further experimental work is certainly needed for reliable approaches to these dependencies.

APPLICATION OF THE MODEL TO THE MTO-PROCESS

The present model was applied to the catalytic conversion of methanol to olefins (MTO) which is currently investigated by the authors. A circulating fluidized bed reactor having the same geometry as the cold flow model was erected. Investigations on the reaction kinetics were carried out and on the basis of the reaction model by SEDRAN et al. [28]. The kinetic scheme depicted in Figure 8 was developed using the method of lumping the different species to a small number of groups. The parameters of the model have been estimated on the basis of experimental data obtained from a bubbling fluidized bed and a Berty-type fixed bed reactor [29].

The circulating fluidized bed model was applied to this reaction system. Calculations were performed to predict the behavior of the erected reactor. In Figure 9 the example of a model calculation is shown. In this simulation the mass flux of methanol was assumed to be 1.5 kg/h corresponding to a superficial velocity at the reactor base of 0.95 m/s (temperature: 500 °C). The catalyst inventory of the riser was set to 1.0 kg resulting in a bottom zone height of 2.5 m. The plot shows the dependence of the different concentrations on height above the distributor.

As can be seen from Figure 9 the concentration profiles exhibit discontinuities at the splash zone location. The reason for this phenomenon is the mixing condition that is assumed to characterize the splash zone. In Figure 10 the predicted axial concentration profile of the educt, methanol, is shown for all phases. It can be seen, that high differences between the phases at the splash zone level exist. Thus, a mixing condition will result in a sharp change of the concentrations. Since discontinuities are unlikely to give a good description of the reactor behaviour further investigations should focus on better splash zone modelling.

Though the solids concentration in the bottom zone is high and should therefore cause a high chemical conversion, the gradient of the calculated methanol concentration is not higher than above the splash zone. Figure 10 shows that the reason for this finding is the mass transfer limitation in the bottom zone of the circulating fluidized bed that results in a high concentration difference between the bubbles and the suspension phase. These model predictions lead to the conclusion, that the catalyst efficiency in the recirculation zone is better than in the bottom zone.

The accuracy of the present model can only be evaluated by comparing the predictions to experimental data. Measurements of the actual MTO-reactor performance are currently carried out and will be published elsewhere [29]. The direct comparison between model and experiment will then also include a discussion of how the underlying assumptions of the present model effect its accuracy.

CONCLUSIONS

- A reaction model for the circulating fluidized bed reactor is proposed that accounts for the observed axial nonuniformities and backmixing phenomena.
- Mass exchange parameters were determined by separate tracer gas experiments.
- As an example for application of the model, it was used to predict the performance of a MTO reactor.

ACKNOWLEDGEMENTS

Financial support by the Deutsche Forschungsgemeinschaft through the Sonderforschungsbereich 238 is gratefully appreciated. In addition, the authors would like to express their thanks to Arne Rosenhagen and Michael Loss for their valuable experimental work.

NOTATION

a	[m²/m³]	interfacial area based on reactor volume
A_t	[m²]	cross sectional area of reactor
c_V	[-]	solids volume fraction
c_W	[-]	drag coefficient
C_i	[mol/m³]	concentration of gas component i
d_p	[m]	Sauter diameter
d_t	[m]	tube diameter
ε	[-]	voidage
ε_b	[-]	bubble fraction
fd	[-]	average fraction of cross-sectional area occupied by the dense phase
fdr	[-]	local fraction of cross-sectional area occupied by the dense phase
F	[kg/h]	feed
g	[m/s²]	gravity acceleration
G_s	[kg/m²s]	solids mass flux based on cross sectional area
H	[m]	height of reactor
H_b	[m]	height of bottom zone
H_s	[m]	height of splash zone
H_r	[m]	height of recirculation zone
h	[m]	height above distributor
k	[m/s]	mass transfer coefficient
k_j	[m³/s kg]	kinetic parameter of reaction j
k_j	[m⁶/s kg mol]	kinetic parameter of reaction j, second order
k_{ig}	[m/s]	mass transfer coefficient bubbles/suspension, component i
K_Q	[1/s]	convective transport bubbles/suspension
L_{pro}	[m]	Length of suction probe
M	[kg]	catalyst inventory of riser
n	[-]	exponent of the RICHARDSON and ZAKI equation, or number of gas components
P	[N/m²]	pressure
r	[-]	dimensionless distance from the riser's centerline
R	[m]	radius of riser
R_i	[mol/kg s]	reaction rate for component, i, based on catalyst mass
Re_t	[-]	Reynolds number based on terminal settling velocity of single particle
$\Re$	[J/mol K]	gas constant

ρ_s	[kg/m³]	solids density
t	[s]	time
T	[°C]	temperature
u	[m/s]	superficial velocity
u_O	[m/s]	superficial velocity at the riser entry
u_d	[m/s]	interstitial gas velocity in the dense phase
u_l	[m/s]	interstitial gas velocity in the lean phase
u_{mf}	[m/s]	minimum fluidization velocity
u_s	[m/s]	superficial gas velocity in the suspension phase
u_t	[m/s]	single particle's terminal settling velocity
v	[m/s]	solids velocity
$\dot{V}$	[m³/s]	volumetric flow
$\dot{V}_{conv}$	[m³/s m]	convective flux between dense and dilute phase per unit height
$\dot{V}'$	[m³/s]	volumetric flow, upper splash zone boundary

subscripts

b	bubble phase
$conv$	convective exchange
d	dense phase
f	feed
i	component i
j	reaction j
l	lean phase
mf	minimum fluidization
p	product
pro	suction probe
r	recirculation zone
s	suspension phase, solids
sz	splash zone
t	tube
0	reactor base

LITERATURE CITED

1. Gierlich, H.-H.; Dolkemeyer, W.; Avidan, A., Thiagarajan, N., Chem.-Ing.-Tech., **58**, p. 238-239, (1986).
2. Contractor, R.M.; Chaouki, J., *Circulating fluidized bed as a catalytic reactor*, in: *Circulating fluidized bed technology* III (editors: Basu. P; Horio, M.; Hasatani, M.), (1990).
3. Lin, S.O.; Potter, O.E., Powder Technology, **74**, p. 73, (1993).
4. Van Swaaij, W.P.M., *Chemical reaction engineering review* (editors: Luss, D.; Weekman jr.,V.W.), ACS Washington, p. 329, (1978).
5. Yerushalmi, J.; Avidan, A.A., *High velocity fluidization*, in: *Fluidization* (editors: Davidson, J.F.; Clift, R.; Harrison, D.), Academic Press, London, p. 284, (1985).
6. Edwards, M.; Avidan, A., Chem. Eng. Sci., **41**, p. 829, (1986).
7. Li, J.; Weinstein, H., Chem. Eng. Sci., **44**, p. 1697, (1989).
8. Kunii, D.; Levenspiel, O., *Fluidization Engineering*, Butterworth-Heinemann, p. 201-206, (1991).
9. Rhodes, M.J.; Geldart, D., Chem. Eng. Res. Des., **67**, p. 20, (1989).
10. Adams, C.K.: *Gas mixing in fast fluidized bed*, in: *Circulating Fluidized Bed Technology II* (editors: BASU, P.; LARGE, J.F.), Pergamon Press, p. 299, (1988).
11. Bader, R.; Findlay, J.; Knowlton, T.M.: *Gas-solids flow patterns in a 12 inch diameter circulating fluidized bed*, in: *Circulating Fluidized Bed Technology II* (editors: Basu, P.; Large, J.F.), Pergamon Press, Oxford, p. 123, (1988).
12. Kagawa, H.; Mineo, H.; Yamazaki, R.; Yoshida, K., *A Gas-Solid Contacting Model for Fast Fluidized Bed*, in: *Circulating Fluidized Bed Technology III* (editors: Basu, P.; Horio, M.; Hasatani, M.), Pergamon Press, p. 551, (1991).
13. Pugsley, T.S.; Patience, G.S.; Berruti, F.; Chaouki, J., Ind. Eng. Chem. Res., **31**, p. 2652, (1992).

14. Li, Y.; Kwauk, M.: *The dynamics of fast fluidization*, in: *Fluidization* (editors: Grace, J.R.; Matsen, J.M.), Plenum Press, p. 537, (1980).
15. Ishii, H.; Nakajima, T.; Horio, M., Chem. Engng. Japan, **22**, p. 484, (1989).
16. Svensson, A.; Johnsson, F.; Leckner, B., *Fluid-dynamics of the bottom bed of circulating fluidized bed boilers*, in: *Fluidized bed combustion XII* (editor: Rubow, L.N.), **2**, p. 887, (1993).
17. Werther, J. Wein, J., *Expansion behavior of gas fluidized beds in the turbulent regime*, AIChE Symposium Series, (1994).
18. Sit, S.P.; Grace, J.R., Chem. Eng. Sci., **33**, p. 1115, (1978).
19. Werdermann, C., *Feststoffbewegung und Wärmeübergang in Zirkulierenden Wirbelschichten von Kohlekraftwerken*, Dissertation, Technical University Hamburg-Harburg, (1992).
20. Brauer, H., *Grundlagen der Einphasen- und Mehrphasenströmungen*, Sauerländer Verlag, Frankfurt, (1971).
21. Richardson, J.F.; Zaki, W.N., Trans. Instn. Chem. Engrs., **32**, p. 35, (1954).
22. Dieterich, E.; Sorescu, G.; Eigenberger, G., Chem.-Ing.-Tech., **64**, p. 136, (1992).
23. Lister, M., *Die numerische Integration hyperbolischer partieller Differentialgleichungen durch die Charakteristikenmethode*, in: Mathematische Methoden für Digitalrechner (editors: Ralston, A.; Wilf, S.H.), Oldenbourg Verlag München, (1972).
24. Rensner, D.; Werther, J., *Estimation of the Effective Measuring Volume of Single-Fibre Reflection Probes for Solid Volume Concentration Measurements.*, Part. Syst. Charact. 10, p. 48, (1993).
25. Zhang, W., Tung, Y., Johnsson, F., Chem. Eng. Sci., **46**, p. 3045, (1991).
26. Werther, J., Chem. Eng. Sci., **47**, p. 2547, (1992).
27. Nelder, J.A.; Mead, R., Comp. J., **7**, p. 308 (1965).
28. Sedran, U.; Mahay, A.; de Lasa, H.I., Chem. Eng. Sci., **45**, p. 1161, (1990).
29. Schoenfelder, H., J. Hinderer, J. Werther, F. Keil, *Methanol to olefins, prediction of the performance of a circulating fluidized bed reactor on the basis of kinetic experiments in a fixed bed reactor*, accepted for publication at the 13th International Symposium on Chemical Reaction Engineering (ISCRE), Baltimore, Sept. 25-28, (1994).

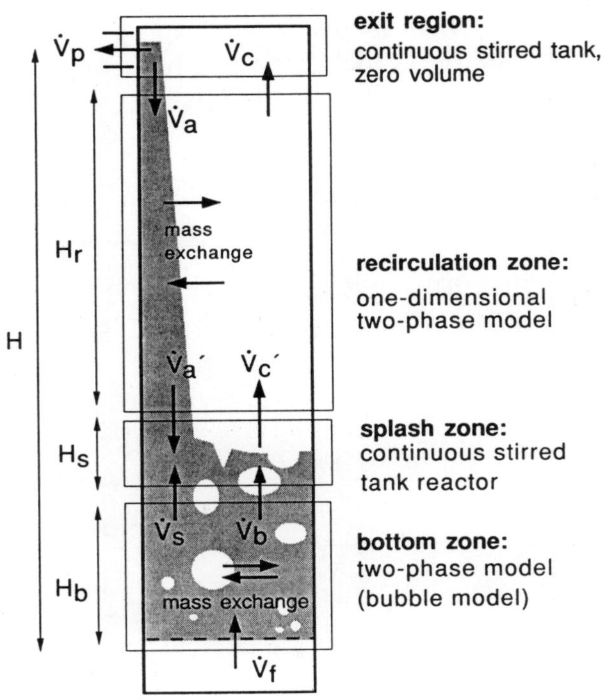

Figure 1. Four stage model of the circulating fluidized bed riser

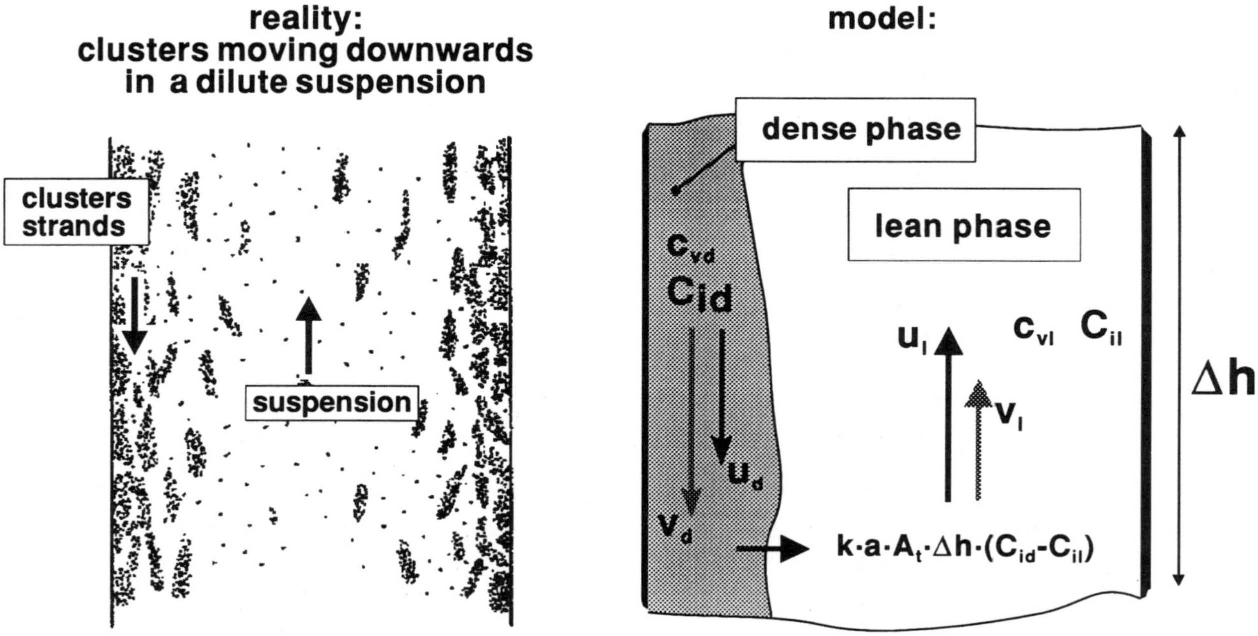

Figure 2. Two-phase counter-current flow modelling approach for the recirculation zone of the circulating fluidized bed riser

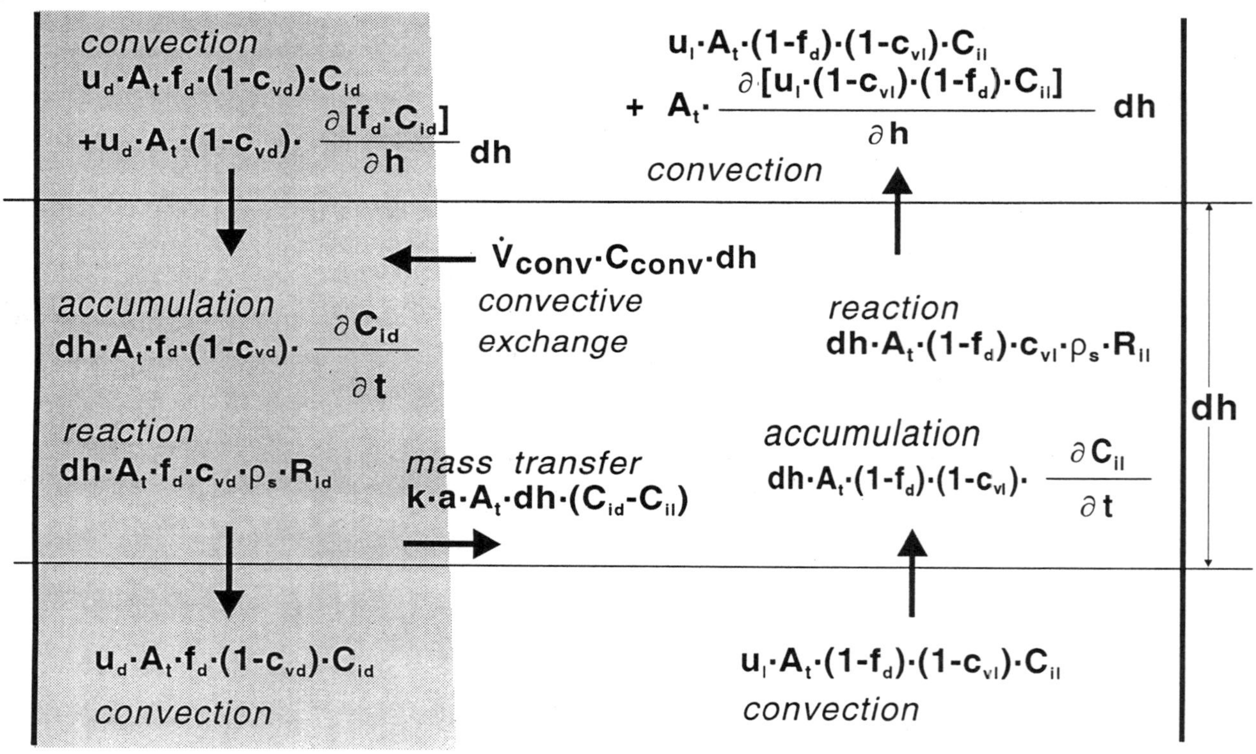

Figure 3. Differential balance for the gas component, i, in the recirculation zone of the circulating fluidized bed riser

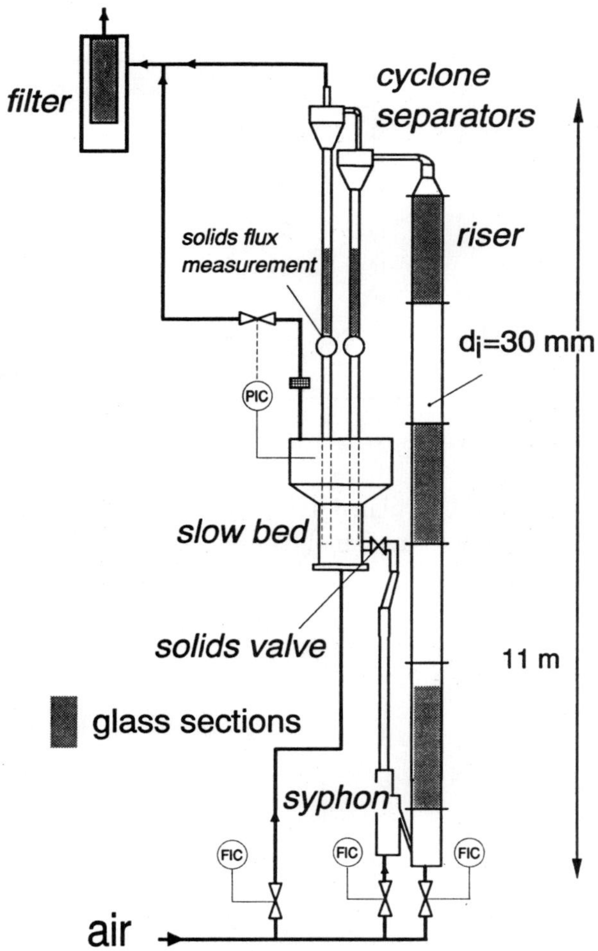

Figure 4. The cold model circulating fluidized bed

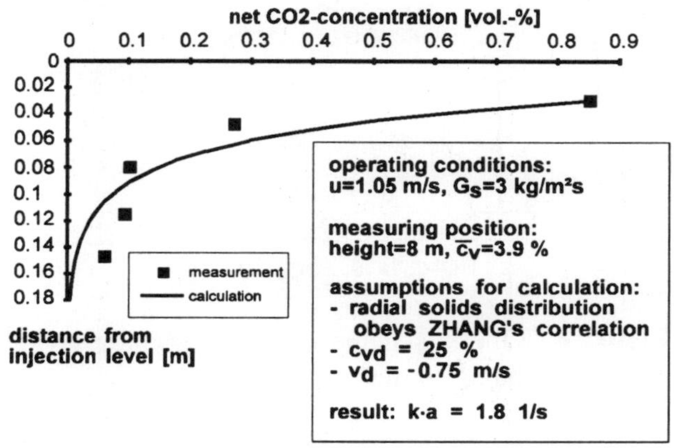

Figure 6. Example of axial backmixing profile for sand particles and CO_2 tracer gas (net concentrations: content of CO_2 in air has been subtracted)

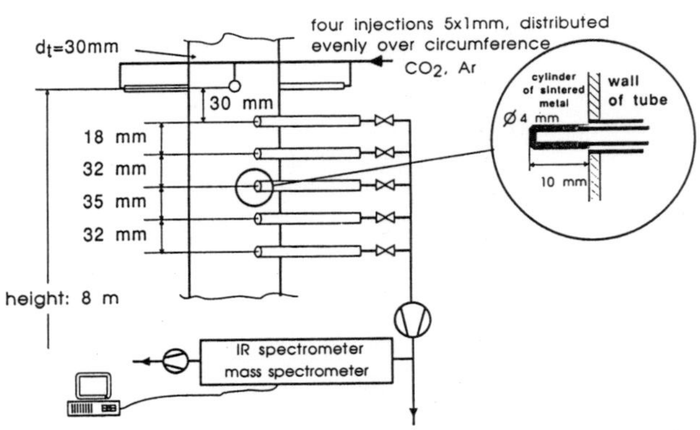

Figure 5. Experimental setup for tracer gas backmixing experiments

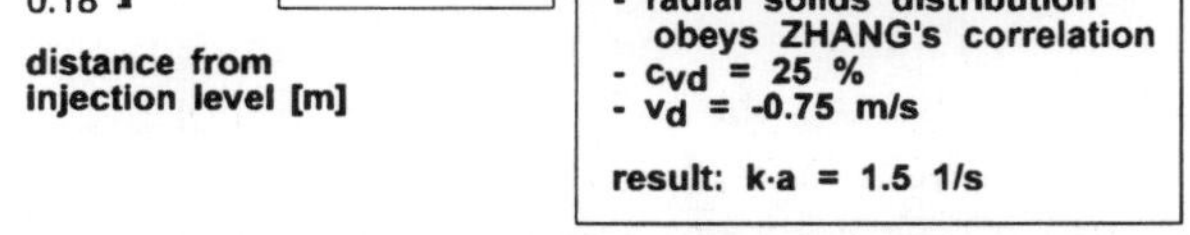

Figure 7. Example of axial backmixing profile for catalyst particles and argon tracer gas (net concentrations: content of argon in air has been subtracted)

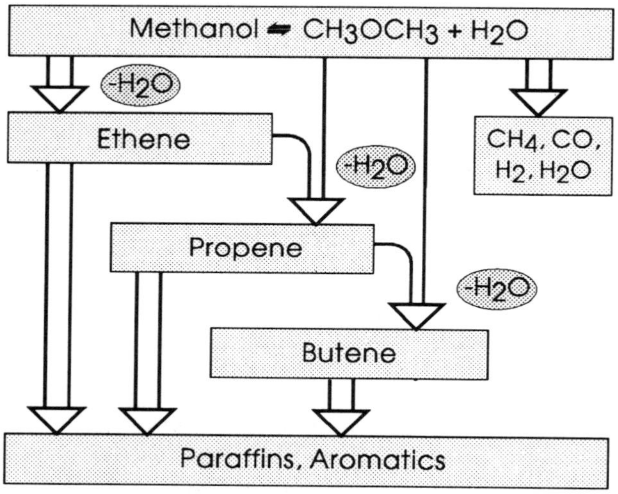

Figure 8. Kinetic scheme of the catalytic conversion of methanol to olefins

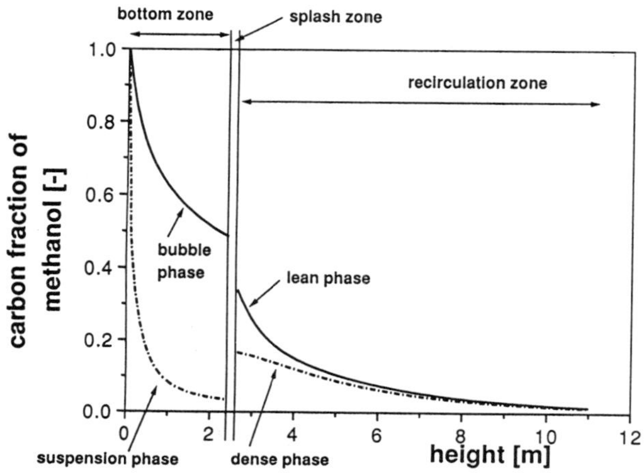

Figure 10. Calculated axial concentration profiles of the educt, methanol (same conditions as in Figure 9)

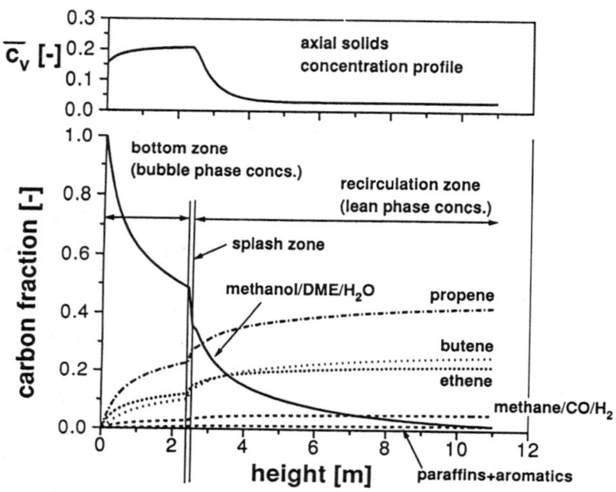

Figure 9. Example of the predicted solids concentration profile and the MTO reactor product distribution in bubble and lean phases, respectively. (T=500 °C, F=1,5 kg/h, u_0=0.95 m/s, M=1 kg)

Radial Profiles of Slip Velocity Between Gas and Solids in Circulating Fluidized Beds

H. Tanner
Swiss Federal Institute of Technology, Institute of Process and Cryogenic Engineering
CH-8092 Zürich, Switzerland

J. Li
Multi-Phase Reaction Laboratory, Institute of Chemical Metallurgy
Academia Sinica, Beijing 100080, P.R. China

and

L. Reh
Swiss Federal Institute of Technology, Institute of Process and Cryogenic Engineering
CH-8092 Zürich, Switzerland

Radial profiles of slip velocity between gas and solids in the riser of a CFB have been determined by measuring the profiles of solids velocity and concentration and calculating the corresponding gas velocity profiles. The measurements have been carried out on a loop seal type CFB with a riser height of 8.5m and an inner diameter of 0.411m. The algorithm to calculate the gas velocity profiles is based on the Principle of Energy Minimization (Li et al., 1988). Within the experimentally covered range of voidages between 0.85 and 1, the local slip velocity increases with decreasing voidage. In the center of the riser, with voidages above 0.98 and particles homogeneously distributed in the fluid, the slip velocity approaches single particle terminal velocity. In the wall region, with local voidages between 0.85 and 0.9 and considerable particle aggregation, the real slip velocity exceeds the terminal velocity of a single sphere under the investigated conditions by up to four times.

The local slip velocity between gas and solids in CFB risers is the key parameter for the description of the fluiddynamics (e.g. momentum balance), the heat and mass transfer between gas and particles, the solids holdup as well as the rate of the mainly diffusion controlled chemical gas/solid reactions (Figure 1). Therefore the knowledge of local slip velocity is crucial to CFB reactor optimization as well as to reliably model the above mentioned phenomena in order to save time and costs with the introduction of new CFB processes or with the scale up of CFB reactors. Today, the measurement of local instantaneous slip velocity is limited to voidages higher than 0.98 (JIN et al. [1]) because adequate measurement systems are lacking.

In this study, the Energy-Minimization-Multi-Scale model introduced by LI et al. [2] was modified in order to calculate radial profiles of gas velocity corresponding to measured particle velocity- and voidage profiles. As a result, radial profiles of slip velocity between gas and solids are obtained.

EXPERIMENTAL EQUIPMENT

Radial profiles of solids velocity and voidage have been measured on a wall-heated, low temperature semi-industrial size CFB (Figure 2). Air as fluidizing agent is supplied by Roots-type blowers.

A narrow sized fraction of spherical glass beads with a mean diameter of $d_{50,3}$=110μm (Ar=99), a density of ρ_p=2500kg/m^3 and a voidage at minimum fluidization of ε_{mf}=0.4 has been used. Without air supply, the bulk inventory in the riser reached a level of 1300mm above the grid.

Superficial gas velocities of up to 6.5m/s were adjusted.

Pressure transducers are installed along the riser height at a distance of 0.7m in order to collect axial pressure profiles.

Supplying heat to the CFB-riser (Figure 3) enables to continuously determine external solids circulation rates by means of an energy balance across the water-cooled siphon in the recycle circuit. Comparative measurements have been performed discontinuously by closing a disk-valve in the downcomer for a short time, preventing the solids to recycle into the loop seal. So, the solids inventory in the downcomer decreases continuously. The solids mass flux is obtained by measuring the change in pressure drop between the bottom of the downcomer and the valve over the time [3].

Radial voidage profiles at different riser levels have been measured by light-backscattering fibre-optical probes with a defined measuring volume (Figure 4, [4]). Radial profiles of solids velocity have been measured with the experimental setup shown in Figure 5. At eight radial position, fifty measurements of instantaneous particle velocity and one hundred measurements of instantaneous voidage were taken and averaged over time.

THEORY

In 1988, Li et al. presented the Energy-Minimization-Multi-Scale (EMMS) model [2]. Physically, it consists of a dense phase of clusters dispersed in a dilute phase of discretely suspended particles in the fluid phase.

Variables are velocities and voidages of gas and solids in dense and dilute phase, the volume fraction of the dilute phase (1-f) and the cluster diameter ℓ as shown in Figure 6.

With respect to the aggregation tendency of gas/solid flow, the model considers different scales of interaction between the two phases (Figure 7):

- micro-scale of particle size
- meso-scale of cluster size
- macro-scale for the overall two-phase flow system including interaction with the boundaries (e.g. walls, internal heat exchangers)

The fluiddynamic equations for the local gas/solid flow are shown below.

Mass balances:

Solids: $c = c_{di}\ (1\text{-}f) + c_{de}\ f$

Fluid: $u = u_{di}\ (1\text{-}f) + u_{de}\ f$

Momentum balances:

Clusters including all dense Phase: $n_{cl} F_{D,cl} + n_{de} F_{D,de} f = n_{cl} F_{G,cl}$

Fluid: $F_{D,di} = F_{G,di}$

Parallel flow of dilute phase and clusters including dense phase:

$$\Delta p_{di} + \Delta p_{cl} = \Delta p_{de}$$

Cluster diameter (LI, 1987):

$$\ell = f\ (P_{Vol}, c, d_p, u_{mf}, \varepsilon_{max}, \varepsilon_{mf}, g)$$

These equations are not sufficient to describe the heterogeneous gas/solid system completely. The concept of minimum potential energy yields an additional constraint: In vertical gas/solid flow, the particles arrange in order to minimize the power needed by the gas for suspending and transporting them (Figure 8).

Mathematically, the model becomes a nonlinear optimization problem: Minimization of P_{Vol} or P_M under the fluiddynamic constraints shown above (P_{Vol} = power/unit volume, P_M = power/mass of solids).

$$P_M = P_{Vol} / ((1\text{-}\varepsilon)\ \rho_p)$$

As a result, it yields the variable vector X:
$X=\{c_{de}, c_{di}, u_{de}, u_{di}, \varepsilon_{de}, \varepsilon_{di}, f, \ell\}$.

Radial profiles of gas velocity

Looking at a particular level of the CFB riser (Figure 9), the power consumption of the gas for suspending and transporting particles must - according to the EMMS - be minimal over the whole cross section.

Assuming an axisymmetric gas/solid flow, this means that for the calculation of radial profiles of gas velocity, the cross-sectional average power to suspend and transport the solids has to be minimized under the constraint that at each radial position also a local minimum for $P_{Vol}(r)$ or $P_M(r)$ respectively is found [5].

Mathematically, the general cross-sectional model yields a twofold nonlinear optimization problem (Figure 10). Because of the high sensibility to input parameters, for enabling calculate radial profiles of slip velocity, the model was simplified as follows:

① Local model:

Simplifications:

- dense phase voidage
  $\varepsilon_{de} = 0.5$
- dilute phase voidage
  $\varepsilon_{di} = 1$
- cluster diameter
  $\ell = f(\varepsilon(r))$

Thus the variable vector X becomes:

$$X = \{u_{de}, u_{di}, c_{de}, f\}$$

Minimizing the power per unit mass of solids for suspension and transport yields:

minimize P_M

$$P_M = \frac{1}{1-\varepsilon} \cdot \frac{1}{\rho_p} \cdot \frac{n_{de} F_{D,de}\, f\, u_{de} + n_{cl} F_{D,cl}\,(1-f)\, u_{di}}{V}$$

This nonlinear optimization problem is solved under the following constraints:

- for the mass balances:

solids: $\tilde{c} = c_{de}\, f$

gas: $\tilde{u} = u_{di}\,(1\text{-}f) + u_{de}\, f$

- for the momentum balance:

clusters including all solids:

$$n_{de} F_{D,de}\, f + n_{cl} F_{D,cl} = (\rho_p - \rho_g)\,(1\text{-}\varepsilon_{de})\, f\ g$$

- Parallel flow of dilute phase and clusters including all dense phase require equal pressure drop:
  $n_{cl} F_{D,cl} / (1-f) = n_{de} F_{D,de}$

- Equality of experimentally obtained and calculated voidage:
  $\tilde{\varepsilon} = \varepsilon_{di} (1-f) + \varepsilon_{de} f$

- gas velocity always exceeds solids velocity and velocity of clusters:
  $u_{slip,de} > 0$
  $u_{slip,cl} > 0$

② Cross-sectional model:
Assumptions:

- boundary conditions (wall influences) are negligible
- properties of gas and solids are constant over the whole cross-sectional riser area
- the cluster diameter $\ell(r)$ can be expressed as a polynomial function of the local voidage $\varepsilon(r)$

Input data:

$\tilde{c}$ (r)	radial profile of solids velocity
$\tilde{\varepsilon}$ (r)	radial voidage profile
$\tilde{u}(r) = \dot{V}_g / A$	global superficial gas velocity
$\rho_g, \nu_g, \rho_p, d_p$	properties of gas and solids
d_r	diameter of CFB riser

Minimization of the power needed to suspend and transport the solids yields:

$$\overline{P}_M = \frac{2}{R^2(1-\overline{\varepsilon})} \int_0^R P_M(r)\,(1-\varepsilon(r))\, r\, dr$$

with $$\overline{\varepsilon} = \frac{2}{R^2} \int_0^R \varepsilon(r)\, r\, dr$$

Constraints:

- at each radial position, a local minimum for $P_M(r)$ must be found
- the measured superficial gas velocity must correspond to the average gas velocity obtained from the calculated gas velocity profile, i.e.

$$\tilde{u} = \dot{V}_g / A = \frac{2}{R^2} \int_0^R u(r)\, r\, dr$$

- the cluster diameter converges:
  for $\varepsilon \to 1$, $\ell \to d_p$
  for $\varepsilon \to \varepsilon_{mf}$, $\ell \to d_r$

RESULTS AND DISCUSSION

The axial voidage profiles in Figure 12 were obtained by differentiating and smoothing the measured pressure profiles in Figure 11.

The radial profiles of voidage and solids velocity (Figure 13) were measured under the operating conditions shown in Figure 11 and at the riser levels indicated in Figure 12. These radial profiles were used as input data for the computation of the radial gas velocity profiles (Figure 10) without being smoothed before.

Also plotted in Figure 13 are the calculated gas velocity profiles over the dimensionless square of the radius.

The relation between the radial profiles of voidage and slip velocity can be found in Figure 14. Within the experimentally covered range of operating conditions and at the investigated riser levels, voidage in the center of the riser generally approaches unity. Solids are homogeneously distributed and the slip velocity approaches its minimum - the terminal velocity of a single sphere. In the wall region, solids aggregation becomes significant with increasing solids concentration. The slip velocity increases to up to four times the terminal velocity (Figure 14 (IV)).

For a single sphere with density of ρ_p=2500kg/m^3 and diameter of d_p=110µm, which corresponds to the mean diameter $d_{50,3}$ of the solids mass fraction used for the experiments, the terminal velocity was calculated to c_t=0.61m/s. Fig 14 (I) shows a nearly uniform slip velocity over the whole radius, which is remarkably smaller than the terminal velocity of a single sphere. This error occurs because the assumption of an axisymmetric flow field is no longer justified near the outlet of the riser into the cyclone. Brereton [6] investigated geometric effects leading to deviations of axisymmetric flow in detail.

The accuracy of the experimentally obtained radial profiles was checked in the following way: From local solids velocity and local solids concentration, the local solids mass flux was calculated. Assuming axisymmetric flow, $\overline{G}_s$ was obtained by integrating it over the coss-section of the riser and dividing it by the cross-sectional area:

$$\overline{G}_s = \frac{2\pi}{A} \int_0^R \tilde{c}(r)\,(1-\tilde{\varepsilon}(r))\,\rho_p\, r\, dr$$

The comparison with $\overline{G}_s$ obtained from disk valve measurements is shown in Figure 15. Calculated and measured solids mass fluxes agree satisfactorily, comparing for example points (II) and (IV), where solids velocity- and voidage profiles have been measured at different riser levels (Figure 12) under constant operating conditions. Experimentally obtained and calculated $\overline{G}_s$ do generally not match when the radial profiles were measured at a level of 6178mm (close to the asymmetric riser outlet) because the assumption of an axisymmetric flow field is no longer justified as mentioned above.

CONCLUSIONS

1) Although the proposed one-dimensional EMMS-model for the calculation of radial profiles of gas velocity had to be simplified in order to stabilize the optimization routine, the calculated gas velocity profiles as well as the obtained slip velocity profiles appear reasonable. Because of lack of adequate equipment for direct gas velocity respectively slip velocity measurement, the results cannot be confirmed by experiments yet.

2) In the proposed model, the drag force acting on particles and clusters was formulated by the correlation of Richardson and Zaki [7]:

$$C_D(\varepsilon) = C_{D0}\, \varepsilon^{-4.7}$$

This correlation is only valid for homogeneously dispersed solids. Müller [8] however proved experimentally that the drag coefficient approaches zero with extreme solids aggregation. Therefore further development of the model must concentrate on finding a more accurate expression for the drag force coefficient as a function of voidage and solids aggregation. A new method to describe clusters in gas/solid flow is presented in [9].

3) Experimental work has to concentrate on the development of a system to measure local voidage, gas- and solids velocity simultaneously. A basic approach in this direction is under elaboration. The direct measurement of solids and gas momentum will avoid statistical problems when measured velocities and voidage have to be combined in order to solve the fluiddynamic equations (e.g. momentum balances) implemented in computer models.

ACKNOWLEDGEMENT

The authors thank the Swiss National Energy Research Foundation (NEFF) for its financial support.

NOTATION

A	Area	[m^2]
c	solids velocity	[m/s]
C_D	drag coefficient	[-]
C_{D0}	drag coefficient for a single sphere	[-]
d,D	diameter	[m]
dp	particle diameter	[μm]
$d_{50,3}$	particle diameter (median of volumetric solids distribution)	[μm]
F	force	[N]
f	volumetric fraction of dense phase	[-]
g	gravity	[m^2/s]
G_s	solids mass flux	[$kg/(sm^2)$]
h,H	height	[m]
ℓ	(hydraulic) cluster diameter	[m]
n	number of...	[-]
p	pressure	[bar]
P	power	[W]
P_M	power per unit mass of solids	[W/kg_p]
P_{Vol}	power per unit volume	[W/m^3]
r, R	radius	[m]
u	velocity, gas velocity	[m/s]
$\dot{V}$	volume flow	[m^3/s]
X	variable vector	[-]

Greek Symbols:

ρ	density	[kg/m^3]
ε	voidage	[-]
ν	cinematic viscosity	[m^2/s]

Indices:

cl	clusters
D	drag
de	dense phase
di	dilute phase
dv	disc valve
G	gravity
g	gas
mf	minimum fluidization
p	particle
r	riser
slip	relative velocity between gas and solids
t	terminal velocity of a single sphere
total	in the whole control volume

Superscripts:

-	cross-sectional average
~	measured value

Dimensionless numbers:

Ar	Archimedes number
Nu	Nusselt number
Pr	Prandtl number
Re_{cl}	cluster Reynolds number
Re_p	particle Reynolds number
Sc	Schmidt number
Sh	Sherwood number

LITERATURE CITED

[1] Yang, Y.L., Jin, Y., Yu, Z.-Q., Wang, Z.-W., "Investigation on Slip Velocity Distribution in the Riser of Dilute Circulating Fluidized Bed", *Powder Technology,* **73**, p67 (1992)

[2] Li, J., Tung, Y., Kwauk, M., "Method of Energy Minimization in Multi-Scale Modeling of Particle-Fluid Two-Phase Flow", *Circulating Fluidized Bed Technology II,* Pergamon Press, p89 (1988)

[3] Goedicke, F., "Strömungsmechanik und Wärmeübergang in zirkulierenden Wirbelschichten", *Dissertation Diss. ETH. Nr. 9733,* (Zürich) (1992)

[4] Reh, L., Li, J., "Measurement of Voidage in Fluidized Beds by Optical Probes", *Circulating Fluidized Bed Technology III,* Pergamon Press, p163 (1991)

[5] Li, J., Reh, L., Kwauk, M., "Application of the Principle of Energy Minimization to the Fluid Dynamics of Circulating Fluidized Beds", *Circulating Fluidized Bed Technology III,* Pergamon Press, p105 (1991)

[6] Brereton, C., "Fluid Mechanics of High Velocity Fluidized Beds", *Ph.D. Thesis, Dept. Chem. Eng., University of British Columbia,* (1987)

[7] Richardson, J., Zaki, W., "Sedimentation and Fluidization", *Trans. Inst. Chem. Eng.* **32** p35 (1954)

[8] Müller, P.B., "Partikel-Widerstand und Druckabfall in beschleunigter Gas-Feststoff-Strömung", *Dissertation Diss. ETH. Nr. 9842,* (Zürich) (1992)

[9] Tanner, H., "Lokale Strömungsmechanik in hochexpandierten zirkulierenden Gas/Feststoff-Wirbelschichten", *Dissertation, Diss. ETH.,* (Zürich) (1994)

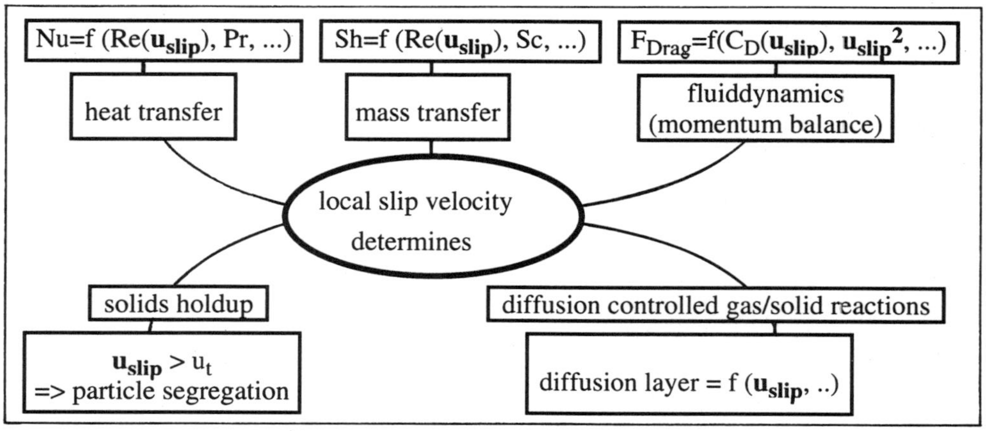

Figure 1: Slip velocity in its context

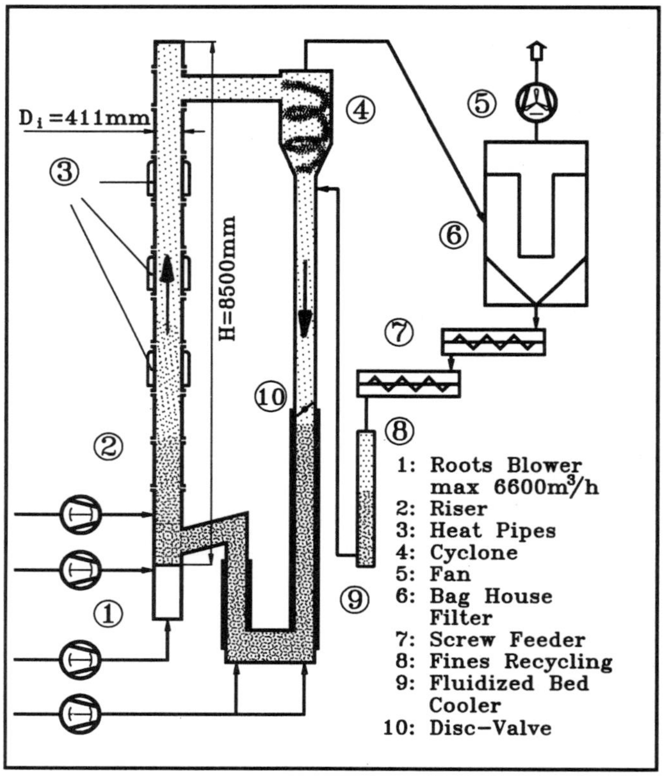

Figure 2: CFB cold model pilot plant

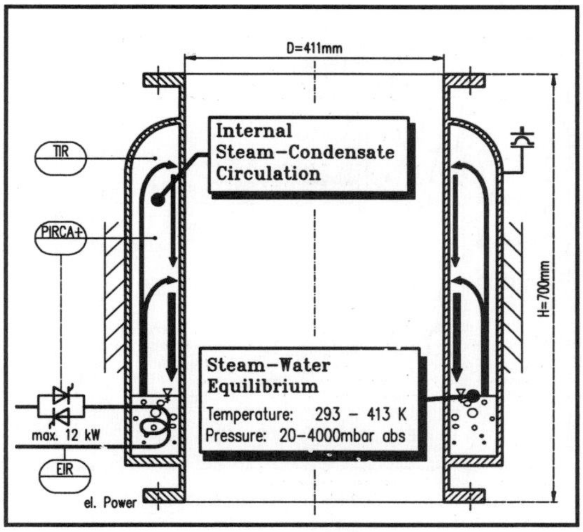

Figure 3: Heat-Pipe-Calorimeter

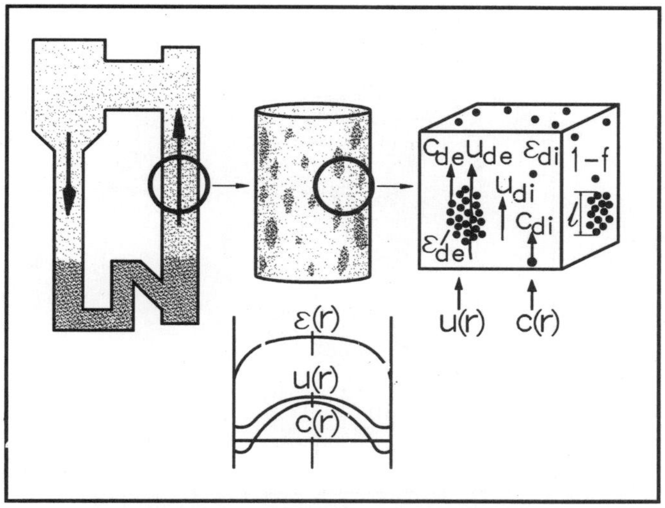

Figure 6: EMMS: Local flow model

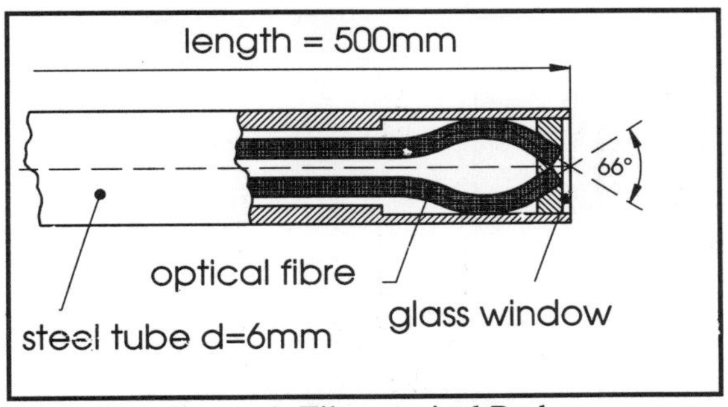

Figure 4: Fiber-optical Probe

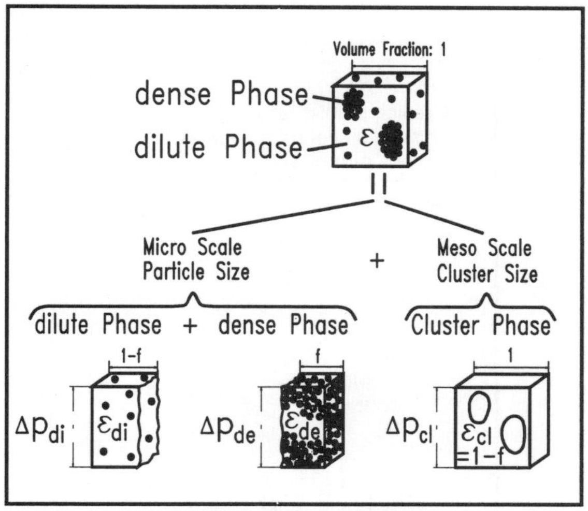

Figure 7: EMMS: Micro- and meso-scale

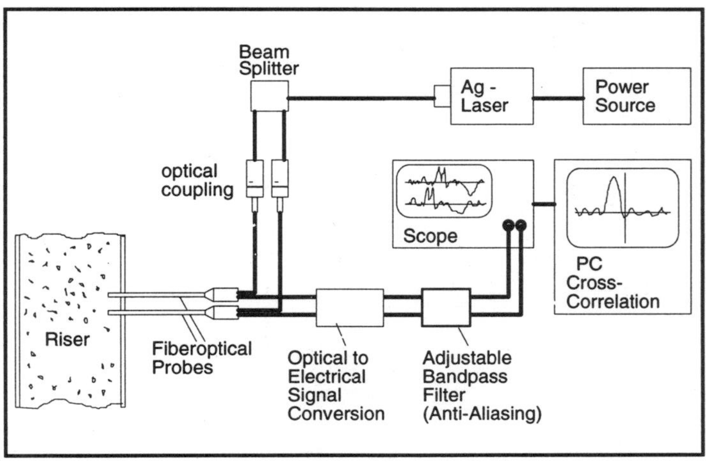

Figure 5: Velocity measurement system

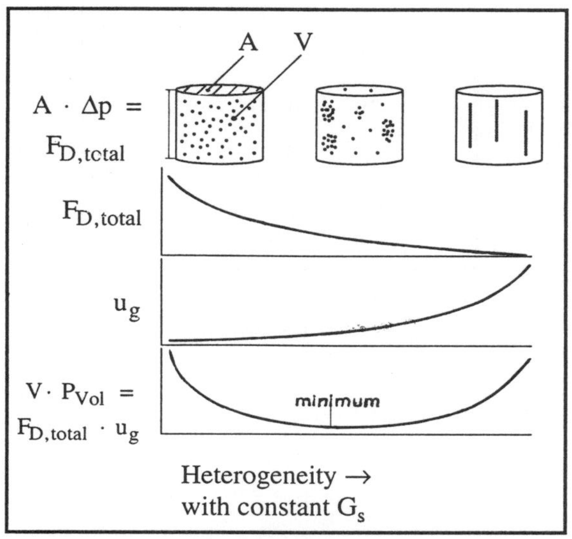

Figure 8: Concept of energy minimization

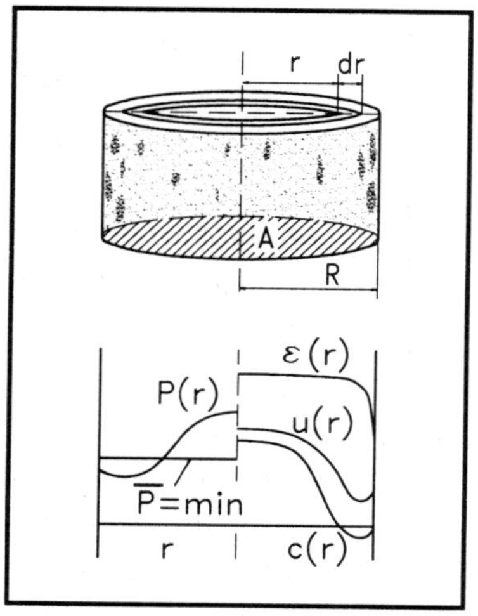

Figure 9: Cross- sectional energy minimization

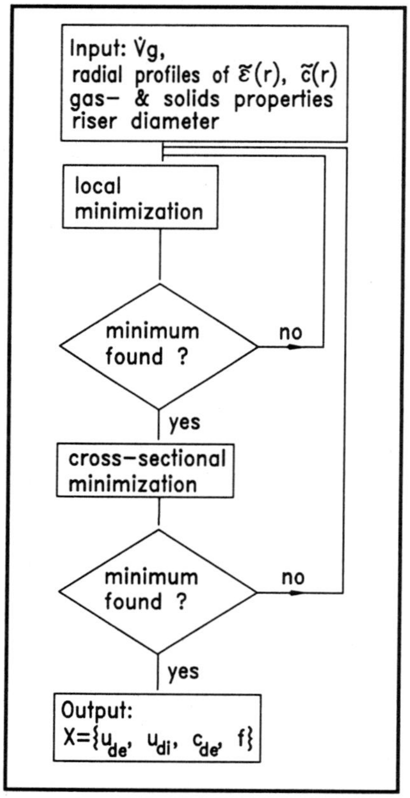

Figure 10: Flow diagram for the cross-sectional model

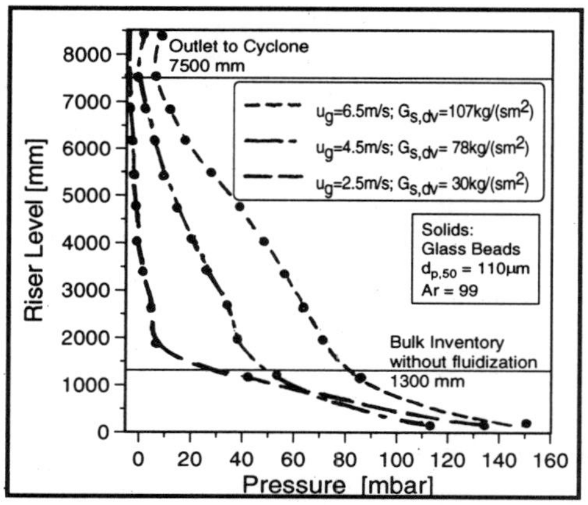

Figure 11: Axial pressure profiles

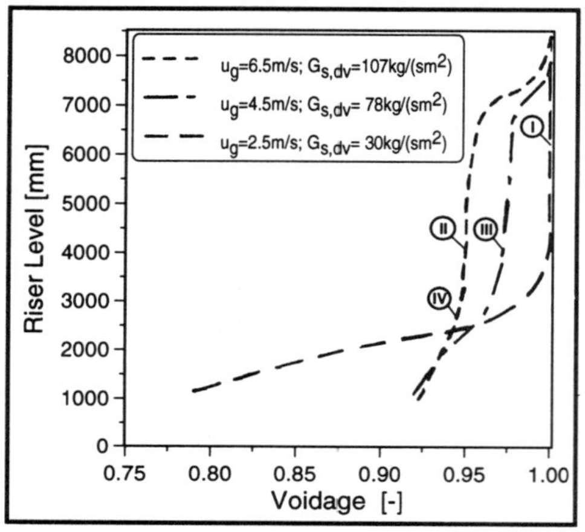

Figure 12: Axial voidage profiles

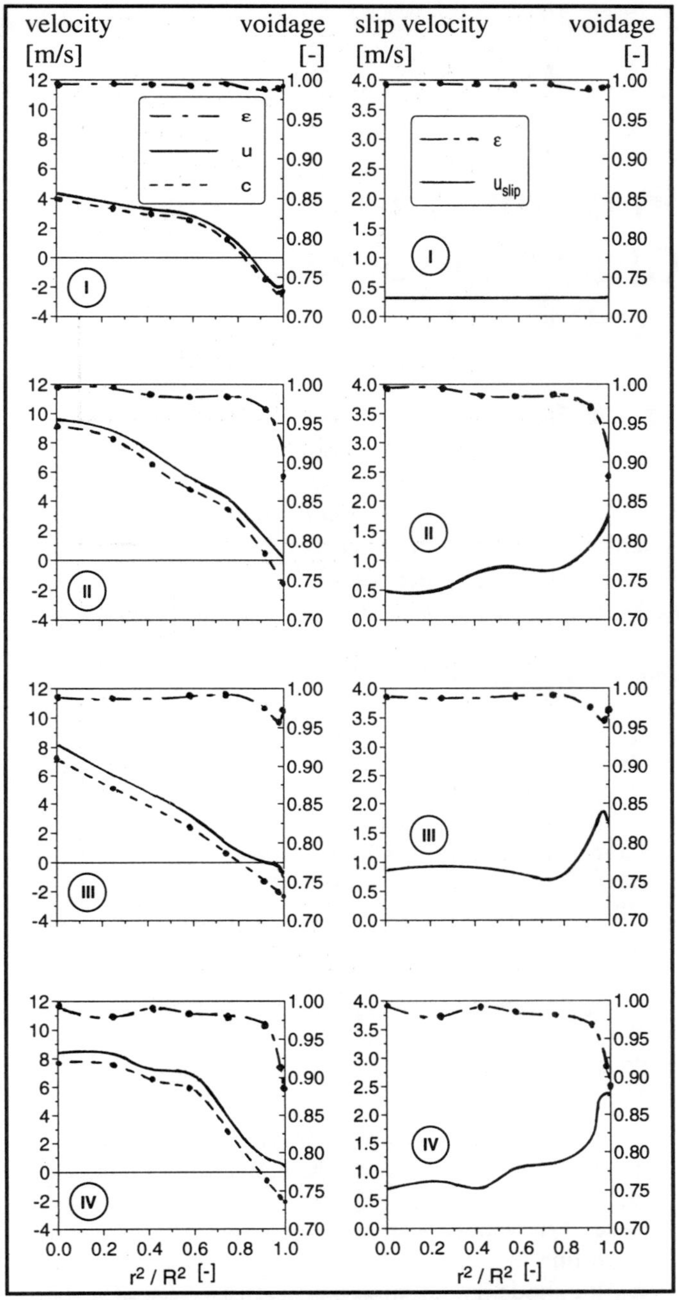

Figure 13: Radial profiles of voidage ε, gas velocity u and solids velocity c

Figure14: Radial profiles of voidage ε and slip velocity u_{slip}

Operating conditions and measurement levels c.f. Figs. 11 and 12

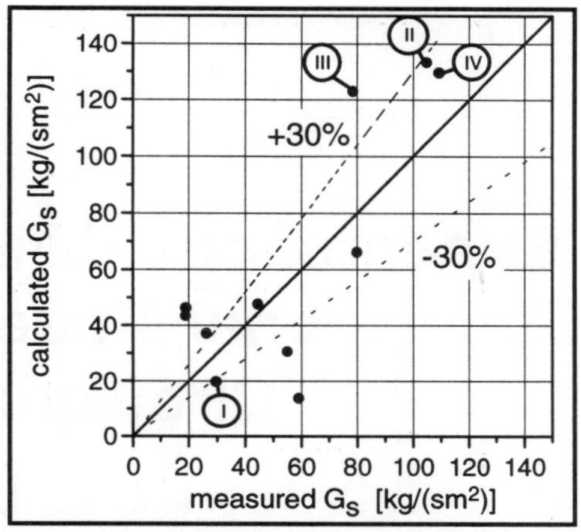

Figure 15: Comparison of measured and calculated solids mass flux

Experimental Validation of an One and Half Dimensional Model of Char Combustion in a Circulating Fluidized Bed

P. Basu, J. Talukdar and Song Wu
Technical University of Nova Scotia, Halifax B3J 2X4, Canada

A mathematical model has been developed for the prediction of combustion of char in a circulating fluidized bed (CFB) furnace. The model predicts the axial distribution of gas concentration, carbon concentration and temperature profile. A core-annulus hydrodynamic model has been used, considering bilateral gas and solid transfer between the core and the annulus, to determine the hydrodynamic parameters. The model has been validated against experimental data generated in a 0.2m x 0.2m x6.5m circulating fluidized bed furnace fired by a bituminous coal. A good agreement between model predictions and experimental data is found..

INTRODUCTION

A mathematical model promotes understanding of a process, and provides valuable information for process optimization, parameter variation, control strategies and scale up consideration. To establish a mathematical model, some fundamental information of the process is required. Although the development and commercialization of circulating fluidized bed (CFB) combustion technology have been extraordinarily rapid through the last decade, a basic understanding of the processes occurring in CFB combustors is still incomplete.

A number of one dimensional models [1, 2 , 3] provided an important first step in the development of a predictive model of CFB furnace. These models, however, overlooked the well recognized core-annulus structure of the CFB furnace. Furthermore, except Arena et al (1991)[3] none offered experimental validation of their model. Hyppanen et al [4], [5] attempted to develop a three dimensional model using their experimental data of solid flow profile as an input to the model. This restricts the general applicability of this model especially for cases where experimental data are not available. Das and Bhattacharya, [6] modified the model of Basu et al [1] introducing core-annulus representation as the basis of their model. The model, however, neglects combustion in the annular region

Thus no comprehensive mathematical model of char combustion is available at present representing the current understanding of the hydrodynamic structure of CFB. The objective of the present work is to develop a model of char combustion in a CFB furnace considering the core-annulus hydrodynamic structure of the furnace, and to verify it against results from a pilot plant test.

THE MODEL

The model is developed considering various physical processes occurring during the combustion of char in a in CFB furnace. The flow structure of a CFB furnace as used in the model is shown in Figure 1. The processes occurring in a circulating fluidized bed are interrelated. So the model is developed on the basis of both energy and mass balance for each component and their simultaneous solution. This is done by dividing the furnace into a number of fully mixed cells. In a CFB combustion system the major part of the combustion and related processes take place in the furnace. The contributions of cyclone and return leg to the combustion and heat transfer process are negligible. So the model is based on the processes occurring in the furnace only. Depending on the gas velocity and solid concentration the lower and the upper part of the furnace behave differently. So it is divided into two

Song Wu is presently at Tampella Power Corporation, 2600 Reach Road, P.O. Box 3308, Williamsport, PA 17701.

main vertical sections: the turbulent bed section which is below the secondary air injection level and the fast bed section which is above this level. Major assumption used in the model are

i. Each cell behaves as a fully mixed reactor.
ii. Bed behaves like a turbulent bed below the secondary air injection level.
iii. Solid flows upwards as well as downwards direction in the turbulent bed, but the gas flows only in the upward direction.
iv. In the turbulent bed the gas and solids are fully mixed in the lateral direction.
v. Above the secondary air injection level, the bed operates in fast fluidized regime.
vi. The fast bed has two distinct zones in the radial direction: core and annulus.
vii. There is no back flow of solids in the core, and no up flow of solids in the annulus of the fast bed.
viii. The gas velocity in the annulus is negligible.
ix. The annulus region has a constant voidage, but its thickness decreases along the bed height.
x. Segregation of particle in the furnace is negligible.
xi.. Combustion and heat transfer are negligible in the cyclone and the return leg.
xii. Mixing of the feed inside the riser is instantaneous and so is the drying and heating process.
xiii. Primary fragmentation of feed particle (char) is instantaneous.
xiv. The carbon monoxide (CO), formed during char reaction, burns off close to the surface of the char. So the CO concentration in the fully mixed gas phase is negligible.
xv. The temperature of char particles is 100 K higher than the bed temperature.

The model described here considers the vertical as well as radial variation of physical properties. However in the radial direction variation is not continuous. It is limited only to two sections: core and annulus. Strictly speaking, the model does not fit into the definition of either one dimensional or two dimensional representation. So call it 'one and half dimensional' model

Turbulent Bed Section:

Below the secondary air injection level the gas velocity is low, but it is sufficiently high to be in a turbulent bed regime. Gas solids suspension density in this section is uniform. Bed solids are drained continuously from the bottom of this section to maintain same bed inventory. The feed and recycled solids are introduced at the bottom of this section. Mass balance, energy balance and population balance are carried out for each cell of turbulent bed can be described as below.

Mass Balance : Carbon balance :

$$M_{s,i}\frac{dC_{ti}}{dt} = G_{u,nt}C_{t,i-1} - (G_{u,nt} + G_{d,nt} + D_r + B_{t,i})\,C_{t,i} + G_{d,nt}C_{t,i} + R_{cr}C_{cy} + FC_F \qquad (1)$$

Oxygen balance :

$$M_{g,i}\frac{dO_{ti}}{dt} = G_{g,i}O_{t,i-1} - (G_{g,i} + B_{ot,i})\,O_{t,i} + S_{t,i} \qquad (2)$$

Energy Balance:

$$(C_{ps}M_{s,i}+C_{pg}M_{g,i})\frac{dT_{ti}}{dt}=(C_{ps}G_{u,nt}+C_{pg}G_{g,i-1})T_{t,i} + C_{ps}G_{d,nt}T_{t,i+1} - (C_{ps}(G_{u,nt} + G_{d,nt}) + G_{gi}\,C_{pg})T_{t,i} + C_{pf}FT_f + C_{pg}G_{ar}\,T_{ar} + H_{r,i} - H_{l,i} + R_{cr}T_{cy}\,C_{ps} \qquad (3)$$

Population Balance:

Population balance of char particle within the size interval d and d+Δd for any cell gives

$$FC_FX_{F,j} - D_rC_iX_{t,i,i} + R_{cr}X_{cy,j} - \frac{\frac{dd}{dt}}{\Delta d_j}X_{t,i,j-1}M_{s,t,i.}C_{t,i} + G_{u,i-1}\,C_{t,i}\,X_{t,i,j} + \frac{3M_{sti}C_{ti}X_{tij}}{d_j}\frac{dd}{dt} + \frac{\frac{dd}{dt}}{\Delta d_j}X_{t,i,j+1}M_{s,t,i}\,C_{t,i} - (G_{u,i}+B_{t,i})X_{t,i,j}C_{ti}+G_{d,i}C_{t,i+1}X_{t,i+1j} - G_{d,i-1}C_{t,i}\,X_{t,i,j} = 0 \qquad (4)$$

Fast Bed Section:

The region between the secondary injection level and the exit of the furnace, the gas velocity is higher and the bed is in fast fluidization regime. The fast behave differently in radial direction which can be seen from Figure 1. First the fast bed section is divided into a number of vertical section. Each section is divided further into two cells: one for the core and one for annulus (Figure 2). Unlike the turbulent bed section, the fast bed is non-uniform in both axial and lateral directions. The axial voidage profile (cross section average) exponential distribution was given by Kunii and Leveinspeil [7] as,

$$\varepsilon_x = \varepsilon^* + (\varepsilon_o - \varepsilon^*)\exp(-a\,x) \qquad (5)$$

For the 0.3 MWt CFB pilot plant used in this work to generate experimental value of decay constant 'a', the bulk density ρ_b of the riser, is correlated by an empirical expression based on the experimental data [8].

$$\frac{\rho_b}{\rho_s} = 0.02624\left[\frac{Gs}{U\rho_g}\right]^{0.801}\left[\frac{Ut}{U}\right]^{1.5} \quad (6)$$

Integrating equation (5) the overall bed density could be found. The equation (6) also gives the overall bed density. So from Equations (5) and (6) the decay constant 'a' could be found. The following assumptions have been made for defining the flow pattern of the fast bed section.

Solid falls through the stagnant gas film in the annulus at their terminal velocity, U_t. So the downward flow of solids

$$G_{di} = \rho_s (1-\varepsilon_a) U_t(A-A_{c,i}) \quad (7)$$

The upward flow of solids along the core is

$$G_{ui} = \rho_s (1-\varepsilon_c) (U_{gc,i} - U_t)A_{c,i} \quad (8)$$

Equating the average voidage in a section with the voidage in the core and annulus of that section along with its area we can find

$$\frac{A}{A_{ci}} = \frac{\varepsilon_{ci}-\varepsilon_a}{\varepsilon_i-\varepsilon_a} \quad (9)$$

The difference between the upward and downward flow of solids gives the net upward solid flux, known as the circulating rate,

$$GsA = G_{u,i} - G_{d,i} \quad (10)$$

Since the entire gas flows through the core area, $A_{c,i}$, the gas velocity in the core can be written from a mass balance as,

$$U_{gc,i} = \frac{UA}{A_{ci}} \quad (11)$$

The solid and gas flow in core and annulus for any cell is shown in Figure 2.

Solid and Gas Exchange Between Core and Annulus:

Solid and gas exchanges ($E_{s,i}$,$E_{g,i}$) between the core and the annulus region are critical to the combustion and sulphur capture reaction. Bai et al [9] correlated solids exchange coefficient based on their experimental results as

$$\frac{f D_T \rho_g}{\mu} = 7.676\, m^{-0.75} Re^{1.04} Fr^{-0.702} Ar^{-0.486} \quad (12)$$

where f = transport rate of solids from the annulus to the core (m/s)and $m = \frac{\rho_g U_{gc}}{Gs}$

If we consider that 'E_s' is the amount of solid flow rate from the annulus to the core in any section of height Δ H , then

$$E_{s,i} = f(1-\varepsilon_i)\,)\,\rho_s\left[\frac{4A_{ci}}{\pi}\right]^{0.5} \pi\,\Delta H \quad (13)$$

E_s represents the bilateral solid transfer between core and annulus which is equal in both direction. There is also net solid transfer from core to annulus which can be determined from mass balance of solid

$$E_{n,i} = G_{ui-1} - G_{u,i} = G_{d,i-1} - G_{d,i} \quad (14)$$

Lateral diffusion of gas can be correlated from the experimental result of Adam [10] as

$$D_{l,i} = (12.1\ Ugc_i/(G_s\rho_s)+ 7)\ 10^{-4} \quad (15)$$

The gas flow rate, E_g, for a section of height ΔH is given by

$$E_{g,i} = \pi\,(4A_{c,i}/\pi)^{0.5}\,\Delta H\,\rho_g \frac{D_{li}}{\Delta R_i} \quad (16)$$

Applying avove equations for each cell balance equaiton can be described for each cell as follows.

Mass balance for carbon in core:

$$M_{sc,i}\frac{dC_{ci}}{dt} = G_{u,i-1}C_{c,i-1} - (G_{u,i} + E_{s,i} + E_{n,i} + B_{c,i})\,C_{c,i} + E_{s,i}\,C_a \quad (17)$$

Mass balance for carbon in annulus:

$$M_{sa,i}\frac{dC_{ai}}{dt} = G_{d,i}C_{a,i+1} - (G_{d,i-1} + E_{s,i} + B_{a,i})\,C_{a,i} + (E_{s,i}+E_{n,i})C_c \quad (18)$$

Mass balance for oxygen in core:

$$M_{gc,i}\frac{dO_{ci}}{dt} = G_{g,i-1}O_{c,i-1} - (B_{o,ci} + G_{g,i} + E_{g,i})\,O_{c,i} + E_{g,i}\,O_{a,i} + S_{c,i} \quad (19)$$

Mass balance for oxygen in annulus:

$$M_{ga,i}\frac{dO_{ai}}{dt} = (B_{oa,i} + E_{gi})\,O_{a,i} + E_{g,i}\,O_{c,i} \quad (20)$$

Energy balance for core:

$$(C_{ps}M_{s,i}+C_{pg}M_{g,i})\frac{dT_{ci}}{dt} = (C_{ps}G_{u,i-1} + C_{pg}G_{g,i-1})T_{c,i-1} - C_{ps}(G_{u,i}+C_{pg}G_{g,i}+E_{s,i}+E_{ni})T_{c,i} - E_{s,i}T_a + H_{r,i} - H_{l,i} \quad (21)$$

Energy balance for annulus:

$$(C_{ps}M_{s,i}+C_{pg}\,M_{g,i})\frac{dT_{ai}}{dt} = (C_{ps}G_{d,i})T_{a,i+1} - (C_{ps}(\,G_{d,i-1} + E_{s,i})+C_{pg}E_{g,i})T_{a,i} + (C_{pg}E_{g,i}+ C_{ps}(E_{s,i}+E_{n,i})T_c+H_{r,i}-H_{l,i} \quad (22)$$

Population Balance:

Population balance of char particle within the size interval d and d+Δd for any cell gives

Core :

$$\frac{\frac{dd}{dt}}{\Delta d_j}X_{i,j-1}\,M_{c,i}C_{c,i} + G_{u,i-1}C_{c,i}\,X_{c,i,j} + \frac{3M_{ci}C_{ci}X_{cij}}{d_j}\frac{dd}{dt} + \frac{\frac{dd}{dt}}{\Delta d_j}X_{c,i.j}\,M_{c,i.}C_{c,i} - (G_{ui}+E_{s,i}+E_{ni}+B_{c,i})X_{c,i,j}C_{c,i} + E_{s,i}C_aX_{a,ij} = 0 \quad (23)$$

Annulus :

$$\frac{\frac{dd}{dt}}{\Delta d_j}X_{a,i.j-1}\,M_{a,i}C_{a,i} + \frac{3M_{ai}C_{ai}X_{aij}}{d_j}\frac{dd}{dt} + \frac{\frac{dd}{dt}}{\Delta d_j}X_{a,i,j}\,M_{a,i}C_{a,i}\,\frac{dd}{dt}\; -(E_{s,i}+B_{a,i})X_{a,i,j}C_{a,i}+E_{s,i}C_cX_{c,i,j}+G_{d,i}C_{a,i+1}X_{a,i+1j} + E_{s,i} + B_{a,i})C_{a,i} = 0 \quad (24)$$

Cyclone and Return Leg:

Combustion is neglected in the cyclone and return leg. For solid return cyclone grade efficiency is considered from mass balance of particles. and this was measured by Wu [8] for the cyclone of the pilot plant

CHAR COMBUSTION :

The rate of change of char size

$$\frac{d(d)}{dt}_{com} = \frac{2\,K\,O}{\rho_{ch}\,C_{ch}} \quad (25)$$

To account for the both kinetic and mass transfer resistance as the burning rate coefficient is written as

$$K = \frac{1}{\frac{1}{R} + \frac{d}{Sh\,Dg}} \quad (26)$$

The Sherwood number can be found from Basu and Subbarao [11]

$$Sh = 2.0 + 0.6\,(\,Re_T\,)^{0.5}\;Sc^{\,0.33} \quad (27)$$

The surface reaction rate expression was measured for the Prince coal which was used in the pilot plant tests [12]. It is given as

$$R=0.0117\exp\left(\frac{-2859}{T_p}\right)\;:\;T_p > 1000\ K \quad (28)$$

REDUCTION IN PARTICLE SIZE:

Primary Fragmentation:

The size distribution of coal after primary fragmentation has been related to that of the feed [13]

$$X_{F,j} = X_{Fo,j}\,(\,d_jk_F^{1/3}\,)\,k_F^{1/3} \quad (29)$$

Attrition:

The total rate of change of char particles size is the result of both combustion and attrition. Rate of change of size due to attrition

$$\frac{dd}{dt}_{at} = -\frac{k_aU\,d}{3d_{av}} \quad (30)$$

So the rate of change of size of char particles will be written as

$$\frac{dd}{dt} = \frac{dd}{dt}_{at} + \frac{dd}{dt}_{com} \tag{31}$$

Halder and Basu [14] measured attrition rate coefficients of relatively coarse carbon particles in circulating fluidized bed combustors. The value of the attrition rate coefficient was in the range of 3.37×10^{-7} - 5.59×10^{-7} [15].

SOLUTION SCHEMES FOR THE MODEL:

The mass balance, population balance and energy balance equations described earlier are solved under steady state conditions. Each set of balance equations for each species is solved separately. A multi-layer iteration method was tried for solution of the whole sets of equations as a system. It was found successful in terms of convergence speed. The flow diagram of solution procedure is shown in Figure 3.

EXPERIMENTAL SET UP

The experimental set up consists of a furnace (0.2 m x 0.2 m crossection, 6.5 m tall), cyclone, a impact separator and return leg. A schematic of the experimental set up is shown in Figure 4. Inner wall of the furnace, cyclone and return leg are refractory lined and there is no heat transfer surface. Primary air enters the bed through a nozzle type distributor plate. The secondary air is injected radially at a height of 1.5 m above the distributor plate. Fuel is fed to the bed at the bottom part. The solid-return from cyclone is also located near bottom part of the riser. The pressure drop and temperature are measured at different height of the furnace. Carbon samples were taken at different locations. Carbon in the fly ash is calculated from the fly ash analysis. The gas sampling were taken at different location of the furnace and measured continuously.

RESULTS AND DISCUSSION

The experimental condition, operating parameters and fuel properties are given in Table 1. Figure 5 shows the suspension density profile in the fast bed of the riser. Suspension density at the bottom most part agrees well with the model. For the top part (from 2.5 m to 6m) agreements are reasonable. But for the region of height 0.5m to 2.5m experiment value and model does not agree well. This may be because of lower assumed value of voidage in the model for the turbulent section. But at the present state of understanding not much known about the turbulent section of the bed and therefore further research is recommended. However present model was used against a pilot plant test where there is no heat transfer surface. So variation of suspension density between model and test result would not effect the overall performance much.

Figure 6 shows the predicted oxygen profile along the bed height. A considerable difference of oxygen concentration between the core and the annulus exists in the lower part of the fast bed section. This difference decreases towards the top of the combustor. Measured values are close to those predicted for the core by the model.

In Figure 7 the predicted carbon fraction in bed materials is plotted against furnace height. In the fast bed section, the carbon contents of core and annulus close to each other, indicating a strong lateral mixing of bed material. In the axial direction, the carbon content decreases along the bed height. Experimental values of the carbon content, though scattered, shows the similar declining tendency along the bed height.

Figure 8 presents a comparison of the measured and predicted values of bed temperatures along the bed height. Since there is no cooling surface in the combustor, the bed temperature increases with the height. It is also noticed that the secondary air brings the bed temperature down by about 20 °C.

Figure 9 gives the predicted size distribution of char particles in the turbulent bed, the fast bed and the return leg. It is observed that there are more fines in the fast bed section than in the turbulent bed. This is due to the simultaneous burning and combustion assisted attrition of char particles. Also, owing to the elutriation effect of the cyclone, the material in the return leg is coarser than that the top of the fast bed. Since the samples of bed materials taken from the bed are mixtures of char, sorbent and inert particles, and it is difficult to separate them, no data for the char size distribution is available for comparison with the model prediction. Effect of attrition and combustion on size reduction of char was studied using the model. It was found that the reduction of char size due to combine effect of combustion and attrition was about 78% of the feed size. Whereas the reduction of char size due to combustion without attrition was only 7.6%. Reduction of char size only because of attrition (without combustion) was 9.6%.

CONCLUSIONS

Predicted results of the model for oxygen, carbon and temperature profiles have been compared with the experimental results obtained from the 0.3 MW pilot plant. A good agreement between the predicted and experimental data is observed. Predictions of char size distributions in various sections of the CFB system show the turbulent bed to contain coarser char particles and the fast bed, finer particles. This agrees with the physical processes occurring in the furnace.

SYMBOLS AND ABBREVIATIONS:

a = Exponential decay constant of Equation 5.
A = Cross sectional area of the furnace, m^2
Ar = Archemedes number
BC = Carbon consumption in any cell, kg
B_oO = Oxygen consumed in any cell, kg
C_F = Carbon fraction in the coal
C = Carbon concentration in any cell
C_{ps} = Specific heat of solids, kJ/(K.kg)
C_{pg} = Specific heat of gas, kJ/(K.kg)
C_{pf} = Specific heat of fuel fed to the furnace, kJ/(K.kg)
d = Size of char, m
d_{av} = Average size of char in the bed, m
D_g = Molecular diffusitivity of oxygen, m^2/s
D_l = Gas transfer coefficient between core and annulus, m/s
D_r = Bed drain, kg/s
D_T = Hydraulic diameter of the riser, m
E_g = Gas transfer rate between core and annulus, kg/s
E_n = Net solids transfer rate from core to annulus
E_s = Solid transfer rate between core and annlus, kg/s
f = Solid transfer coefficient between core and annulus, m/s
F = Fuel feed rate, kg/s
Fr = Froud number
G_{ar} = Air flow rate to the furnace, kg/s
G_d = Downward solids flow rate, kg/s
G_g = Upward gas flow rate, kg/s
Gs = Solid recirculation flux, kg/(s.m^2)
G_u = Upward solids flow rate , kg/s
H_l = Heat loss , kJ/s
H_r = Heat generated, kJ/s
k_a = Attrition rate constant, kg/(m2.s)
k_F = Fragmentation factor
K = Overall burning rate , kg/(m^2.s)
M_g = Mass of gas at any cell of the bed, kg
M_s = Mass of total solids in any cell of the bed, kg
O = Oxygen fraction in any cell, kg/m^3
R = Surface reaction rate, kg/(m^2.s.(kg O_2/m^3))
Re = Reynold's number
R_{cr} = solid recirculation rate, kg/s
R_{eT} = Reynolds number based on terminal velocity
S = Oxygen coming from primary air and fuel, kg/s
Sc = Schimt number
Sh = Sherwood number
T = Temperature of bed
T_{ar} = Temperature of air coming to the furnace, C
T_f = Temperature of fuel coming to the furnace
T_p = Char particle temperature in the furnace, C
t = Time
U = Superficial gas velocity, m/s
U_{gc} = Gas velocity inside the core, m/s
U_t = Terminal velocity of the particle, m/s
x = Height of the furnace from the distributor plate, m
X_j = Weight fraction of char of size j in any cell
X_{Fj} = Weight fraction of coal entering the furnace
X_{FOj} = Weight fraction of coal entering the furnace
ε, ε_x = Voidage in any section of the furnace.
$\varepsilon*$ = Voidage at the infinite height of the furnace
ε_o = Voidage in dense section of the furnace
ρ_b = Average gas solids suspension density in fast bed, kg/m^3
ρ_s = Density of bed material, kg/m^3
ρ_g = Gas density, kg/m^3
μ = Viscosity of gas, kg/(m.s)
Δd = Size interval of char, m
ΔH = Height of each cell, m
ΔR = Thickness of annulus, m

Subscript

a = Annulus
at = Attrition
c = Core
ch = Char
com = Combustion
cy = Cyclone
i = ith cell
j = Mean size interval
g = Gas
nt = Topmost cell of turbulent bed
s = Bed material (solid)
t = Turbulent bed

REFERENCES

1. Basu, P. , A. Sett and E.A.M. Gbordzoe " A Simplified Model for Combustion of Carbon in a Circulating Fluidized Bed Combustor" in 9th Int. Conf. on Fluidized Bed Combustion. J.P. Mustonen ed. ASME, NY pp. 738-742(1987)

2. Weib, V. , J. Schpler and F.N. Fett " Mathematical Modeling of Coal Combustion in a Circulating Fluidized Bed Reactor," in Circulating Fluidized Bed Technology II, P. Basu and J.F. Large, eds. Pergamon Press, Oxford, pp.289 - 298.(1988)
3. Arena, U., A. Malandrino and L. Massimilla ,"Modeling of Circulating Fluidized Bed Combustion of Char" , Canadian Journal of Chemical Eng., 69 Aug. 1991, pp 860 - 868.(1991)
4. Hyppanen, T., Y. Y. Lee and Raimio " A Three Dimensional Model for Circulating Fluidized Bed Combustion", in Circulating Fluidized Bed Technology III, P. Basu, M. Horio and M. Hasatani, eds. Pergamon Press, Oxford, p.p. 563 -568.(1991)
5. Hyppanen, T. , Y. Y. Lee and Raimio " A Three Dimensional Model for Circulating Fluidized Bed Boiler", FBC Technology ASME, (11 th Conf) , pp 439 - 448.(1991)
6. Das, A. and S. C. Bhattacharya "Circulating Fluidized - Bed Combustion " Applied Energy 37 (1990) pp 227 - 246.(1990)
7. Kunii, D. and O. Levenspiel "Entrainment of Solid for Fluidized Bed ", Powder Technology, 61, pp.193-206., (1990)
8. Wu, S. . "Combustion and Sulphur Capture in Circulating Fluidized Beds", Ph.D Thesis, Technical University of Nova Scotia, Halifax, Canada.(1991)
9. Bai, D., Y. Jin, and Z. Yu, " The Two-Channel Model for Fast Fluidization. " Fluidization'88, M. Kwauk, D. Kunii, eds., Science Press, Beijing, pp.155-164. (1988)
10. Adam, C. K. , "Gas Mixing in Fast Fluidized Bed", in Circulating Fluidized Bed Technology II, P. Basu and J.F. Large, eds. Pergamon Press, Oxford, pp. 299-306.(1988)
11. Basu, P. and D. Subbarao " An Experimental Investigation of Burning Rate and Mass Transfer in Turbulent Fluidized Bed", Combustion and Flame 66: pp. 261-269(1986)
12. Wu, S and P. Basu , "Surface Reaction Rates of Coarse Bituminous Char Particles in the Temperature Range of 600-1340K" Fuel, In Press.(1993)
13. Bellgardt, D.F. , F. Hembach , M. Schessler. and J. Werther , " Modeling of Large Scale Fluidized Bed Combustors", Proceeding of 9th Int. Conf. on Fluidized Bed Combustion, J,P. Mustonen ed., ASME, pp. 713 - 772.(1987)
14. Halder, P. K., P. Basu , " Attrition of Spherical Electrode Carbon Particle during Combustion in a Turbulent Fluidized Bed ", Chemical Engineering Science, Vol. 47, No. 3 pp 527 - 532.(1992)
15. Halder, P. K. (1989) " Combustion of Single Coal Particle in Circulating Fluidized Bed", Ph.D. Thesis, Technical University of Nova Scotia, Canada.

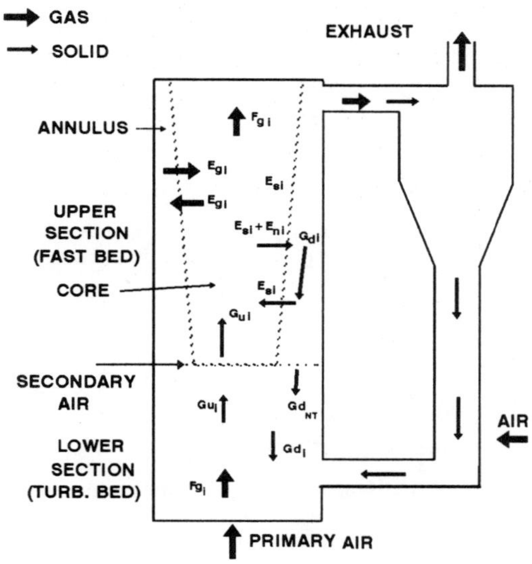

Figure 1. Flow Structure of a CFB

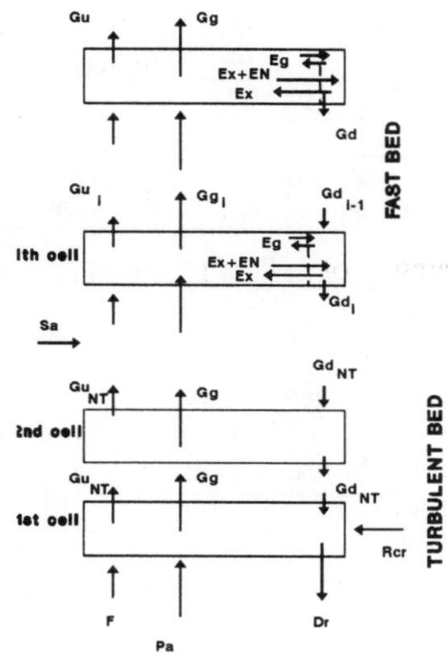

Figure 2. Representation of gas-solid flow in a CFB furnace in cell mode

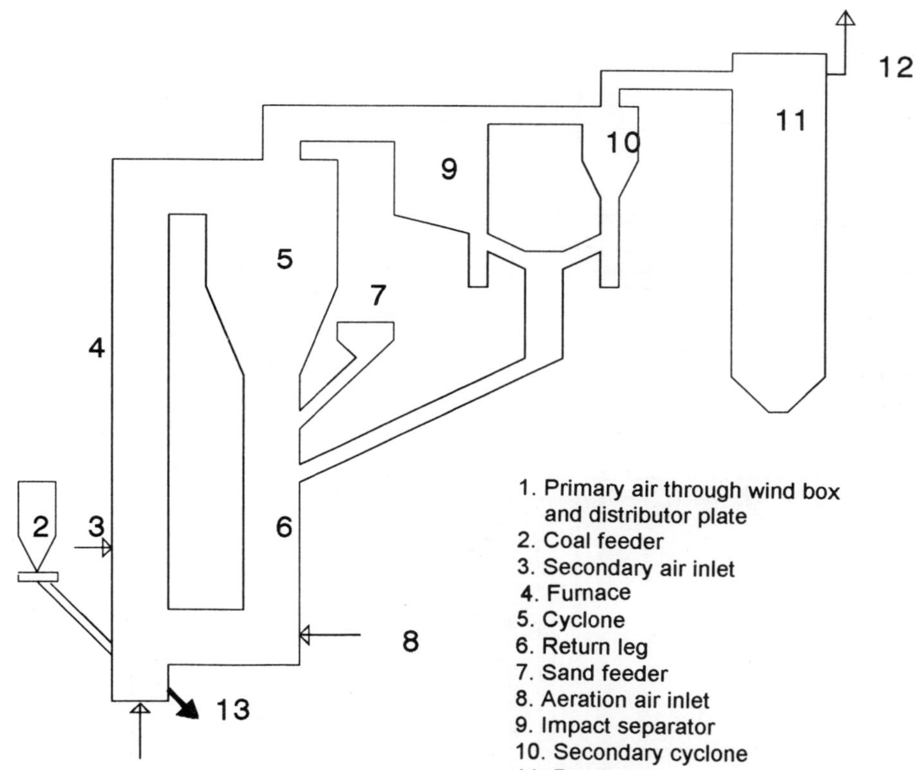

1. Primary air through wind box and distributor plate
2. Coal feeder
3. Secondary air inlet
4. Furnace
5. Cyclone
6. Return leg
7. Sand feeder
8. Aeration air inlet
9. Impact separator
10. Secondary cyclone
11. Bag house
12. Exhaust
13. Bed drain outlet

Figure 4. Schematic of the Pilot Plant

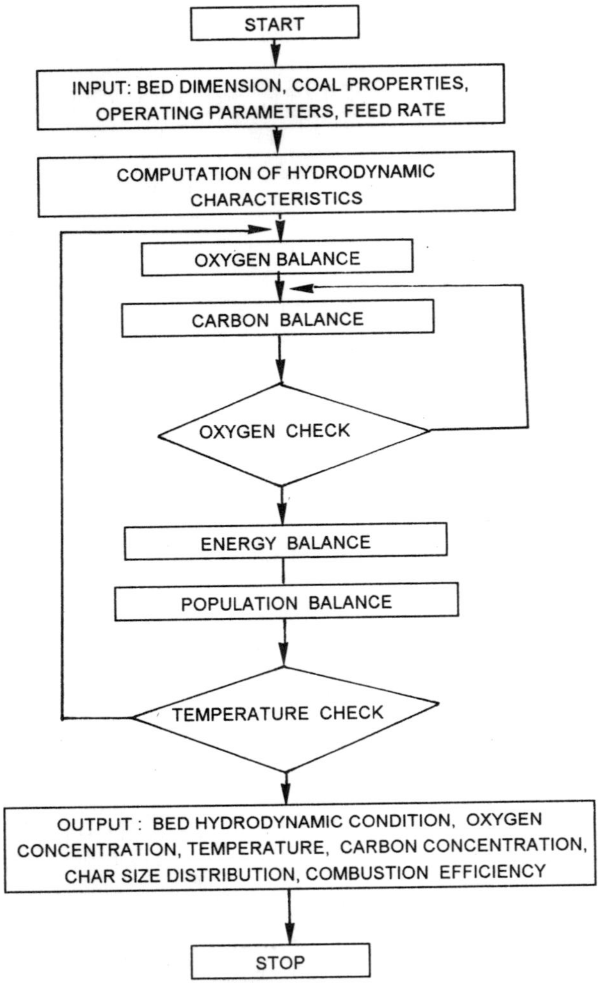

Figure 3. Flow chart for model solution

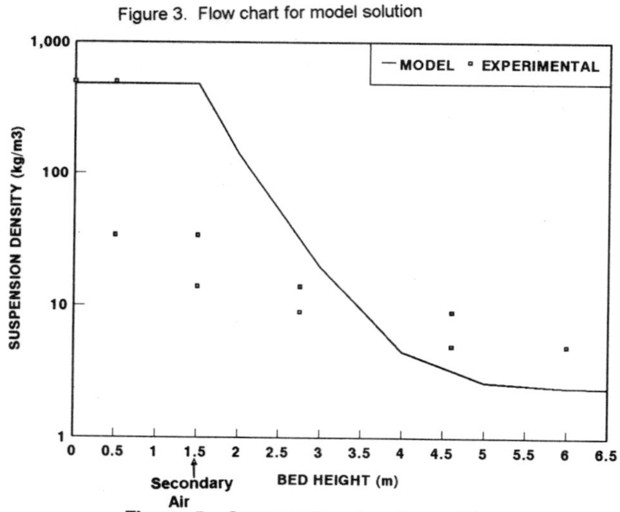

Figure 5. Suspension density profile

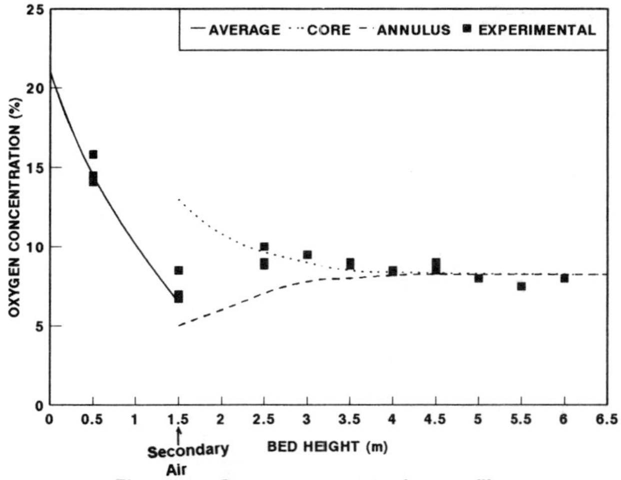

Figure 6. Oxygen concentration profile

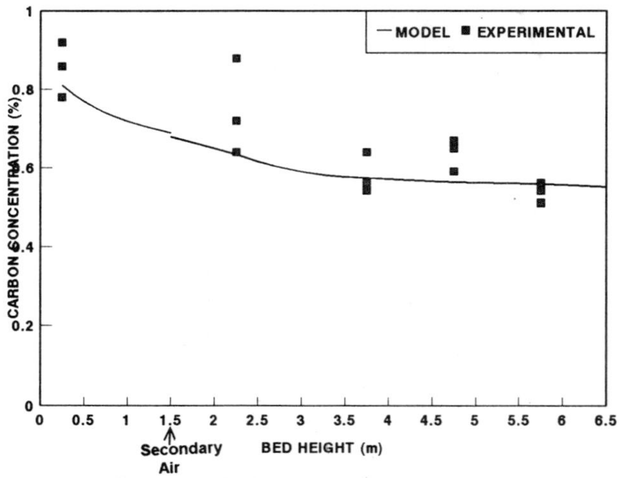

Figure 7. Carbon concentration profile

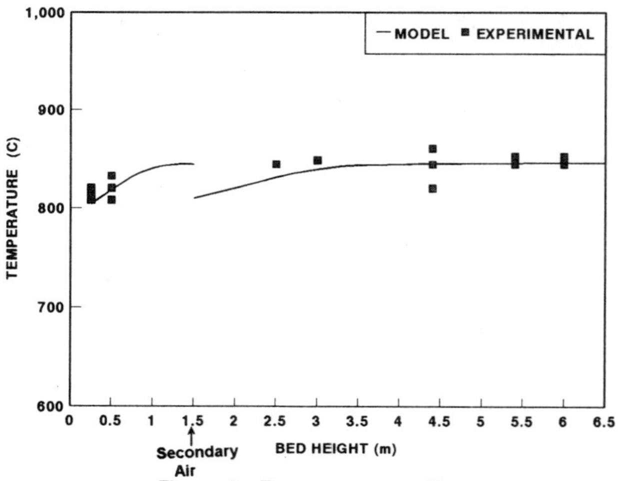

Figure 8. Temperature profile

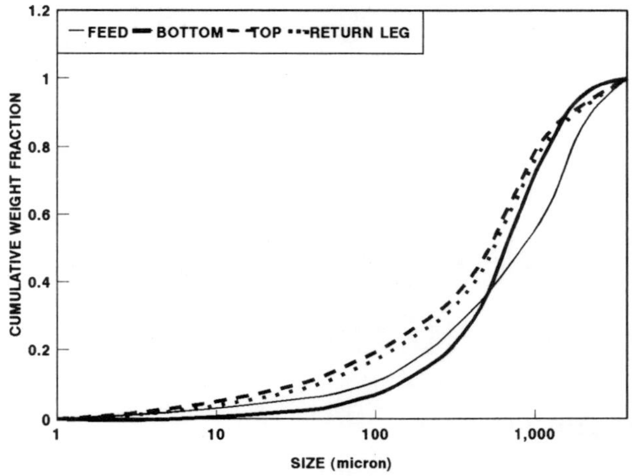

Figure 9. Size distribution profile

Table 1. Model input parameters

Furnace Dimensions	Bed height	6.5 m
	Height below secondary	1.5 m
	air level	0.203 m
	Depth of the furnace	0.203 m
	Width of the furnace	
Feed Condition	Fuel feed rate	0.0077
	Fixed carbon	52.92 %
	Volatile content	30.3 %
	Carbon	69.42 %
	Sulphur	3.57 %
	Moisture	3.57 %
	ash	13.15%
Operating Parameters	Superficial gas velocity at lower section	5.0 m/s
	Superficial gas velocity at upper section	8.0 m/s
	Solid circulation flux	20 (kg/m^2Kg)
Densities	Density of gas	0.32 (kg/m^3)
	Density of bed material	2400 (kg/m^3)

Fluidization of Cohesive Powders in a Sound Waves Vibrated Fluidized Bed

Lii-ping Leu and Chen-Tung Huang
Department of Chemical Engineering National Taiwan University, Taipei 106 Taiwan, R.O.C.

The behavior of eight types of Geldart C particles was studied in the sound waves vibrated fluidized beds (SVFB). A speaker powered by a function generator was used to generate sound as the source of vibration of a fluidized bed. From the measurement of bed pressure drop and bed expansion data of SVFB, the interparticle force and the Group C/A boundary of lower density powders were estimated.

The handling of fine particles (< 30 micron) is very important for various industrial applications and processes because they often have high surface to volume ratio. For example, aerogel powders that have high surface to volume ratio, appear very attractive in chemical reactors. In ceramic industry or powder metallurgy, smaller primary size particle commonly gives a better quality of final product. Furthermore, fine particles are also gaining increasingly importance in the manufacture of drugs, metals, foods, plastics, etc. However, the fluidization of fine particles is generally difficult to achieve.

According to Geldart classification [1], the fine particles (Group C) have strong cohesive force. "Normal" fluidization of such powder is very difficult; the powders lift as a plug in small diameter beds or form cracks, bridges and channels in large diameter beds. These phenomena arise from the fact that the interparticle forces are greater than those which the fluid can exert on the particle. Generally speaking, interparticle forces are the result of van der Waals, electrostatic, capillary forces, wet and sticky particle surface, irregular particle shape, etc.

Fortunately, the fluidization of Group C particles can be improved by mechanically agitation or vibration (Mori et al. [2], Dutta and Dullea [3], Jaraiz et al. [4]), acoustic vibration (Morse [5], Nowak and Hasatani [6,7], Chirone et al. [8,9]) and addition of fluidization aid (Lauga et al. [10], Dutta and Dullea [11]). Vibration overcomes the interparticle forces and breaks up the agglomerates, so cracks, bridges or channels diminish or disappear. Due to more complete gas/solid contacting, higher bed pressure drop and bed expansion are observed.

Baerns [12] studied the feasibility of gaseous fluidization of particles less than 50 micron and defined an index describing the fluidizability of a particulate material. Geldart et al. [13] reported that the relatively small changes in particle size and other parameters which affected interparticle forces can transform a fine free-flowing powder into one that is cohesive. Morooka et al. [14] investigated the fluidity of submicron powders in a small-scale transparent column. They found that Ni, Si_3N_4, SiC, Al_2O_3, and TiO_2 were smoothly fluidized when the gas velocity was increased enough to disintegrate the particle structure and to prevent the coalesce of agglomerates. Iyer and Drzal [15] studied the behavior of cohesive powders in narrow-diameter fluidized beds. They found that Group C powders are very difficult to fluidize and tend to form cracks at lower velocities and cracks/channels at higher velocities thereby precluding any actual fluidization. Pacek and Nienow [16] investigated the fluidization characteristics of a hardmetal powder of extremely high density and small primary particle size. They found the powder can be fluidized via self-agglomeration.

Mori et al. [2] developed a vibro-fluidized bed to fluidize Group C particles and found that a wide range of fine particles down to submicron level can be fluidized fairly well at relatively lower gas velocity. Dutta and Dullea [3] used external vibration to improve the quality of fluidization of a fine cohesive powder by simultaneous increasing the bed pressure drop, the bed expansion, and decreasing the elutriation loss. Jaraiz et al. [4] also obtained the similar results.

Morse [5] found that low frequency (50 to 500 Hz), high intensity(above 110 dB) sonic energy will markedly improve the fluidization of nonfluent, fine-grained powders with channeling and stagnation. Nowak and Hasatani [6] used a speaker powered by an audio amplifier as the source of vibration. They pointed out that the quality of fluidization of cohesive powders is very good (larger bed expansion and higher heat transfer coefficient) when low frequency acoustic energy of sufficient intensity is introduced. Chirone et al. [8] showed that application of acoustic fields to beds of cohesive powder gives place to a bubble-free fluidization typical of Group A non-cohesive powder. Their finding is related to the disaggregation of large clusters into subclusters.

The first study of powder classification according to their fluidization behavior is Geldart [1]. He classified the powders into four groups characterized by density difference and mean particle size. Molerus [17] proposed a non-dimensional group which defined the limit of fluidizaton by considering interparticle cohesion forces and ended with a classification of powders somewhat like Geldart's. Grace [18] pointed out that the boundary between Group C and Group A is strongly influenced by interparticle force, including van der Waals, capillary, electrostatic and magnetic forces as well as variables such as size of surface asperities, hardness of the solid material, humidity, melting or softening point, electrical conductivity and magnetic susceptibility. Jaraiz et al. [4] developed a theory to estimate the particle to particle cohesive force from pressure drop versus expansion data of packed vibrated beds of very fine particles subject to a gentle upflow of gas. They predicted further the Group C/A transition and compared with other results in the literature.

In order to understand the difficulty of fluidization of cohesive powders and then to improve further the quality of fluidization, the primary objective of this study is to measure the bed pressure drop and bed expansion in the sound waves vibrated fluidized bed (SVFB), estimate interparticle force and predict the boundary between Group C and Group A.

THEORETICAL BACKGROUND

Interparticle force

In this analysis, we followed the theory derived by Jaraiz et al. [4] to find the interparticle force from the measurement of pressure drop and bed expansion data.

Transition from Geldart Group C to A powders

We defined a parameter I as the transition number

$$I = \frac{\text{interparticle force at a contact point}}{\text{gravity force}}$$

$$= \frac{F_{pp}}{G_p} \qquad (1)$$

We should consider that the relative importance between interparticle force at a contact point and gravity force could vary with type of particles and fluidizing gas when the boundary between Group C and A was estimated. That is

$$I_{C/A} = f(\rho_p, \rho_g, d_p, g, \mu) \qquad (2)$$

From dimensional analysis, the following relation could be obtained:

$$(1 / I_{C/A})(\frac{\rho_p^2 d_p^3 g}{\mu^2})^{a/2}(\frac{\rho_g^2 d_p^3 g}{\mu^2})^{b/2} = \text{constant} \quad (3)$$

We assumed further a = 2 and b = 2 for simplicity at Group C/A boundary and defined

$$Ar_p = \frac{\rho_p^2 d_p^3 g}{\mu^2} \quad \text{and} \quad Ar_g = \frac{\rho_g^2 d_p^3 g}{\mu^2}$$

So Eq. (3) could be expressed as

$$\frac{I_{C/A}}{Ar_p * Ar_g} = K = \text{constant} \quad (4)$$

First chose a kind of particle on Geldart C/A boundary and calculated the interparticle force, the value of $I_{C/A}$ from Eq. (1) and K value from Eq. (4). Then the Group C/A transition (or $d_{C/A}$) could be estimated.

Expression of sound pressure level

i (sound pressure level, SPL) is commonly defined as [19]

$$i = 20 \log_{10} (P_{se}/ P_{so}) \quad [dB] \quad (5)$$

and the effective sound pressure (measured or desired) can be obtained as follows:

$$P_{se} = P_{so}\ 10^{i/20} \quad [Pa] \quad (6)$$

where P_{so} is the standard reference of sound pressure (= 2×10^{-5} Pa).

EXPERIMENTAL

The fluidized bed was consisted of a 24 mm i.d. x 1 m Plexiglas tube with a cloth distributor. The distributor was clamped by two pieces of stiff wire screen so that it would not vibrate in subsequent experiments. A speaker rated 30 W with audible range frequency 20 - 20 kHz located at the bottom of the unit was used to generate sound as the source of vibration of the fluidized bed, and it was powered by a function generator with square waves. Any trace of humidity of fluidizing air was removed by a fixed bed of silica gel.

The pressure drop across the empty bed and fluidized bed were measured by the pressure transducer.

Flour (high protein), diatomaceous earth, kaolin, calcium carbonate, sand, silicon nitride, aluminum oxide (α phase) and tungsten were used for study in this work. Their physical properties are shown in Table 1.

RESULTS AND DISCUSSION

Determination of optimum frequency and power

Basically the optimum frequency should be determined from the plot of ΔP versus f and H/H_0 versus f (at some constant superficial gas velocity). When the bed has larger ΔP and H/H_0 at some frequency, the quality of fluidization is better. Nevertheless, there are some practical difficulties. The optimal frequency depends on the length of a riser of empty bed; however, after powder was packed into the bed, some wave traveling from dilute medium (air) to dense medium (packed bed or expanded bed) is partially transmitted and partially reflected. When the reflected wave travels from dense medium to dilute medium, part of the wave is transmitted and part is reflected [20]. Hence, the system forms complex interference of sound waves and it is not the results of resonance in the closed tube as simply described by Nowak and Hasatani [6].

The plug could not collapse at some frequencies or they formed tortuous cracks and small plugs even if the plug collapsed. The bed could have large ΔP and H/H_0 because small plugs existed. Therefore, the optimum frequency could not found only from the data of ΔP and H/H_0 alone. The bed had different bed expansions at different superficial gas velocities; in other words, the density of dense medium (W/A*H) and interference of sound waves changed with superficial gas velocity. Here we found the optimal frequency at 260 Hz by trial and errors method. Figure 1 was the distribution of intensity of sound above the distributor for different frequencies when the bed was empty.

Description of experiments

Experiments with flour

(1) Result of static bed height equals 3 cm (Figure 3)
When the air flow was initially turned on and slowly increased, all the fixed bed lifted as a plug up the column with a relatively high pressure drop. Sound was introduced, then the plug collapsed due to vibration and it became tortuous cracked and small plugs. As the superficial gas velocity was increased, the bed would form a spouted circulating bed (SCB) as shown in Figure 2. If the sound energy was stopped at this velocity, the bed produced a large cavity (i.e., channeling) in the center with very low pressure drop. When the gas velocity was increased continuously during vibration, the bed started smooth expansion with increasing the pressure drop. The bed could be completely fluidized at U = 5.263 cm/s with $\Delta P = \Delta P_W$. If the gas velocity was increased further, the elutriation would be observed.

In nonvibrated beds, the minimum fluidization occurred at U = 5.921 cm/s. Beyond this velocity, the difference between vibrated and nonvibrated beds was small; however, the vibrated beds still had a larger value of ΔP and H/H_0.

(2) Result of static bed height equals 5 cm
The experiments were similar to the ones previously described. Fluidization could be achieved at U = 7.111 cm/s with $\Delta P < \Delta P_W$ and it might be due to some stagnant zones around the distributor.

(3) Comparison with other results in the literature
Jaraiz et al. [4] observed the behavior of flour (density = 1.41 g/cm^3, size = 13.4 -16.9 micron) in mechanical vibrated fluidized bed. They found that ΔP and H/H_0 are larger during vibration, that is the better gas-solid contact. Nowak and Hasatani [6] found flour (density = 0.46 g/cm^3, size = 28 micron) can not be fluidized readily due only to air flow. The channels and stagnant zones disappear when acoustic energy of sufficient intensity is introduced, and the quality of fluidization is very good.

Experiments with diatomaceous earth and tungsten

(1) Result of static bed height equals 3 cm
For diatomaceous earth and tungsten, cracks, bridges and channels were formed all the time even during vibration. Increasing gas velocity, the above phenomena still existed. Because channeling also leaded to high value of local air velocity, undesired elutriation loss occurred. In nonvibrated beds, the beds were in still state.

The results were similar whether the static bed height was 3 cm or 5 cm.

(2) Comparison with other results in the literature
Jaraiz et al. [4] observed the behavior of tungsten (density = 14.52 g/cm^3, size = 0.82 - 3.85 micron) in mechanically vibrated fluidized bed. They found tungsten is an awkward material.

Experiments with kaolin

(1) Result of static bed height equals 3 cm (Figure 4)
When the air flow was initially turned on and slowly increased, the fixed bed rose as a plug. After sound was introduced, the plug collapsed because of vibration and agglomerates forming. The distribution of agglomerates due to segregation was found. It was the smaller ones at the top of the bed and larger ones at the bottom of the bed. As gas velocity increased, agglomerates got smaller and started to vibrate. The agglomerates of the surface started to fluidize at U = 4.605 cm/s. When gas velocity was increased further during vibration, agglomerates got smaller, the range of fluidization region increased and elutriation was observed. After U = 11.611 cm/s, ΔP leveling off was the result of counter-balance of increased ΔP due to increase of gas velocity and decreased ΔP due to elutriation. In addition, after the bed surface had been fluidized, the agglomerates still had the possibility to coalesce again, the interparticle forces could not diminish or disappear and ΔP might be larger than ΔP_W.

In nonvibrated beds, there were some smaller cavities at lower gas velocity. Some particles were seen to vibrate in restricted regions at higher gas velocity.

(2) Result of static bed height equals 5 cm
The results were similar to (1) except the surface fluidization occurred at higher gas velocity.

Experiments with calcium carbonate and sand

(1) Result of static bed height equals 3 cm

The fluidization behavior of calcium carbonate and sand was similar to the behavior of kaolin previously described.

(2) Result of static bed height equals 5 cm
The surface could be fluidized at higher velocity but cracks and channels existed all the time at the middle and bottom of the bed during vibration. Pressure across the bed was dependent on the profile which gas passed through.

(3) Comparison with other results in the literature
Morooka et al .[14] found the agglomerates of calcium carbonate (density = 2.71 g/cm^3, size = 0.554 - 3.4 micron) are very cohesive and easily coalesced to form a homogeneous structure, so they are difficult to be fluidized.

Experiments with silicon nitride
(1) Result of static bed height equals 3 cm (Figure 5)
The results before the surface fluidization at U = 2.631 cm/s were similar to the behavior of kaolin. With an increase in gas velocity, the pressure drop across the bed leveled off. It might be due to that number of contact points did not decrease apparently and the fluidized region did not increase further either, particles adhered to the bed wall to form a thin film and only fewer particles were carried out. Similar to kaolin, the pressure drop across the bed could be larger than ΔP_w.

In nonvibrated beds, the surface fluidization occurred at U = 5.263 cm/s. Beyond this velocity, the difference between vibrated and nonvibrated bed was small; however, the vibrated bed still had a larger value of ΔP and H/H_o. For other static bed heights, the results were similar.

(2) Comparison with other results in the literature
Morooka et al. [14] studied the fluidity of silicon nitride (density = 2.91 g/cm^3, size = 0.13 - 0.155 micron) in a small scale bed. When the gas velocity is increased enough to disintegrate the particle structure and to prevent the coalescence of agglomerates, silicon nitride is smoothly fluidized. Jaraiz et al. [4] observed the behavior of silicon nitride (density = 2.63 g/cm^3, size = 0.7 - 2.11 micron) in mechanically vibrated fluidized bed. From their results, the difference between vibrated and nonvibrated bed is small, and indicating the particles are practically in Group A range.

Experiments with aluminum oxide (α-phase)
(1) Result of static bed height equals 3 cm (Figure 6)
When the air was initially turned on and slowly increased, air merely percolated through the void spaces between stationary particles. After sound was introduced, some small particles flowed upward (i.e., segregation). When U = 1.316 cm/s, the bed started to expand. With a further increase in gas velocity, powder bed entered into transition region. A decrease in pressure drop across the bed might due to that the drag forces suddenly unlocked the four fifths upper part of the bed and the interparticle forces disappear. By observation when U = 3.947 cm/s, the four fifths upper part of the bed was fluidized; however, the bottom of the bed consisted of stagnant agglomerates. As gas velocity was increased continuously, the pressure drop across the bed kept constant and elutriation loss was observed. In nonvibrated bed, the four fifths upper part of the bed was fluidized at U = 5.921 cm/s. Since at this velocity, the behavior of both vibrated and nonvibrated bed was similar; however, the vibrated bed still had larger ΔP and H/H_o.

(2) Result of static bed height equals 5 cm
The results were similar to (1) except the fluidization of four fifths upper part of the bed occurred at higher gas velocity.

(3) Comparison with other results in the literature
Baerns et al. [12] studied the feasibility of gaseous fluidization of aluminum oxide (density = 3.89 g/cm^3 , size = 11 - 13 micron). He observed that fluidization can not be achieved because of severe channeling. Morooka et al. [14] studied the fluidity of alumina oxide (density = 3.89 g/cm^3, size = 3.9 - 5.2 micron). Similar to silicon nitride, aluminum oxide can be smoothly fluidized.

Comparison of SVFB between H_o= 3 cm and 5 cm

The initial superficial velocity making the beds rising as a plug (all except aluminum oxide), the minimum fluidization velocity (for flour) or surface starting velocity (for others) at $H_o = 5$ cm was larger than that at $H_o = 3$ cm. This was due to more number of particles and contact points; that is, larger interparticle forces. In addition, because gas-solid contacting at $H_o = 3$ cm was better than that at $H_o = 5$ cm, $\Delta P/\Delta P_w$ and H/H_o at $H_o = 3$ cm were larger. It was quite different from the usual fluidization of Group B powders that the fluidization of Group C powders depended on the static bed height.

Subgroups of Group C powders

From the above experimental observation, we could classified the Group C particles into the following three subgroups.

(1) complete fluidization
This type of particles had large bed expansion and could be fluidized under bubbling conditions similar to Group A. Flour used in this study belonged to this subgroup.

(2) partial fluidization
This type of particles had medium bed expansion and formed agglomerates during vibration. At a certain gas velocity, the bed gave two layers: a stagnant bottom layer with larger agglomerates and a top layer with smaller agglomerates that fluidized. Kaolin, calcium carbonate, sand, silicon nitride, and aluminum oxide were classified into this subgroup.

(3) complete defluidization
This type of particles had enormous cohesive force. Cracks and channels existed all the time and the quality of fluidization could not be improved even during the vibration of sound. Diatomaceous earth and tungsten were classified into this subgroup.

Estimation of interparticle force

From the pressure drop and bed expansion data, the interparticle force could be estimated by the method of Jaraiz et al. [4], the results were shown in Table 2.

Transition from Group C to Group A powder

In Geldart's classification, the boundary between Group C and A is more uncertain since it is based on different experimental methods and setups, and the available data are scare. In order to point out the C/A transition further, we could use the theory of Jaraiz et al. [4] and transition number to estimate the boundary.

Interparticle number Rc
By the method of Jaraiz et al. [4], we could get C/A transition $d_{C/A}$ based on $R_c = 1$. The results were shown in Table 3. Figure 7 showed our predicted data and other results in the literature on Geldart's chart. Due to the mean particle sizes measured by different methods were quite scattering, our prediction was a band region L_1H_1. Compared with J_1K_1 of Jaraiz et al. [4], we found that their data are located on L_1H_1 in the range of low density (< 4 g/cm^3).

C/A transition number $I_{C/A}$
Flour (density = 1.4678 g/cm^3, size = 20 micron) was located at Geldart C/A transition, so we got K by choosing it as the base then used Eqs.(4) and (1) to find $d_{C/A}$. The results were shown in Table 4 and Figure 8. If we put the data of Dutta and Dullea [3] and Jaraiz et al. [4] on our plot, we could get a band RT and it seemed to parallel the Geldart C/A curve. The difference between RT and Geldart C/A could come from the following reasons: we assumed that the shape of particles was spherical and particles contact with each other at only one point, but actually, the shapes of particles are irregular and the contact areas are larger; the mean particle size is different in each measurement.

CONCLUSIONS

The behavior of Group C powders (flour, diatomaceous earth, kaolin, calcium carbonate, sand, silicon nitride, aluminum oxide and tungsten) was studied in the sound waves vibrated fluidized beds (SVFB). The following results were obtained:
1. The sound waves vibrated powder bed had larger pressure drop and bed expansion during vibration of sound, the quality of fluidization was improved.
2. The group C powders used in this study could be classified into complete fluidization, partial fluidization and complete defluidization three subgroups according to their behavior in the vibrated

bed.
3. The behavior of fluidization of Group C powder was dependent on the static bed height. The lower static bed height was; the better quality of fluidization was.
4. Based on proposed transition number $I_{C/A}$, the Group C/A boundary at lower density (< 4 g/cm^3) could be predicted.

ACKNOWLEDGMENT

The authors are grateful to the National Science Council, R. O. C. for financial support.

NOTATION

A : cross-sectional area of bed [m^2]

Ar_g : $\frac{\rho_g^2 d_p^3 g}{\mu^2}$ [-]

Ar_p : $\frac{\rho_p^2 d_p^3 g}{\mu^2}$ [-]

$d_{C/A}$: particle diameter representing the boundary between Group C and A [m]

d_p : particle diameter [m]

$\overline{d_p}$: mean particle diameter [m]

F_{pp} : interparticle force or force between particles at a contact point [N]

G_p : gravitational force of a particle [N]

f : frequency [Hz]

g : acceleration of gravity [m/s^2]

H : height of expanded bed [m]

H_o : height of fixed bed [m]

$I_{C/A}$: C/A transition number [-]

i : sound pressure level (SPL) [dB]

P_{se} : effective value of sound pressure [Pa]

P_{so} : standard reference of sound pressure (= 2 x 10^{-5} Pa) [Pa]

ΔP : pressure drop across the bed [Pa]

ΔP_w : weight of packed powders/cross-sectional area of bed [Pa]

R_c : interparticle number defined by Jaraiz et al. [4] [-]

U : superficial gas velocity [m/s]

U_{mf} : minimum fluidization velocity [m/s]

W : total weight of packed powders [kg]

Greek Symbols

ε_o : fixed bed voidage [-]

μ : gas viscosity [Pa.s]

ρ_g : gas density [kg/m^3]

ρ_p : particle density [kg/m^3]

LITERATURE CITED

1. Geldart, D., *Powder Technol.*, **7**, 285 (1973).

2. Mori, S., A. Yamamoto, S. Iwata, T. Haruta, I. Yamada and E. Mizutani, *AIChE Symp. Ser.* **86,** 276 (1990).

3. Dutta, A. and L. V. Dullea, *AIChE Symp. Ser.* **87**, 281 (1991).

4. Jaraiz, E., S. Kimura and O. Levenspiel, *Powder Technol.*, **72,** 23 (1992).

5. Morse, R. D., *Ind. Eng. Chem.*, **47,** 1170 (1955).

6. Nowak, W. and M. Hasatani, *Proc. Third Asian Conf. on Fluidized Bed and Three-Phase Reactors (edited by H. S. Chun and S. D. Kim),* p.423, Kyongju, Korea (1992).

7. Nowak, W., M. Hasatani and M. Derczynski, *AIChE Symp. Ser.*, **89**, 296 (1993).

8. Chirone, R., L. Massimilla, and S. Russo, *"Fluidization VII"(edited by O. E. Potter and D. J. Nicklin),* p.545, Engineering Foundation, New York, New York (1992).

9. Chirone, R., L. Massimilla, and S. Russo, *Chem. Eng. Sci.*, **48**, 41 (1993).

10. Lauga, C., J. Chaouki, D. Klvana and C. Chavarie, *Powder Technol.*, **65,** 461 (1991).

11. Dutta, A. and L. V. Dullea, *AIChE Symp. Ser.* **86,** 276 (1990).

12. Baerns, M., *Ind. Eng. Chem. Funda.*, **5,** 508 (1966).

13. Geldart, D., N. Harnby and A. C. Wong, *Powder Technol.*, **37,** 25 (1984).

14. Morooka, S., K. Kusakabe, A. Kobata and Y. Kato, *J. Chem. Eng. Japan,* **21,** 41 (1988).

15. Iyer, S. R. and L. T. Drzal, *Powder Technol.*, **57,** 127 (1989).

16. Pacek, A. W. and A. W. Nienow, *Powder Technol.*, **60,** 145 (1990).

17. Molerus, O., *Powder Technol.*, **33**, 81 (1982).

18. Grace, J. R., *Can. J. Chem. Eng.*, **64,** 353 (1986).

19. Richardson, P. D., *Applied Mechanics Reviews,* **20**, 201 (1967).

20. Giancoli, D. C., "Physics for Scientists and Engineers with Modern Physics", chap. 16 - 17, 2nd edi., Prentice-Hall, Englewood Cliffs, New Jersey (1988).

TABLE 1. Particle density and size of the powders used.

POWDER	ρ_p Accupyc* Pycnometer (kg/m^3)	$\bar{d}_p$ Microtrac* (μm)	$\bar{d}_p$ Sedigraph* 5100 (μm)	$\bar{d}_p$ Malvern* 4700 (μm)	$\bar{d}_p$ Galai* CIS-1 (μm)	$\bar{d}_p$ Coulter* Multisizer (μm)	voidage of fixed bed ε_o
Flour (high protein)	1467.8	20.02			21.39		0.5160
Diatomaceous earth (SiO_2)	2298.5	13.09	6.99		8.23		0.8489
Kaolin	2583.8	2.75	1.73		4.65		0.8167
Calcium Carbonate ($CaCO_3$)	2709.5	4.18	2.44		4.63	2.72	0.6941
Sand	2733.7			2.12	3.57	4.99	0.6824
Silicon Nitride (Si_3N_4)	3087.4			0.72			0.7955
Aluminum Oxide (α-Al_2O_3)	3889.2		0.39	0.78			0.6448
Tungsten (W)	18018.6			1.16			0.8269

*The instruments used for the measurement of particle size.

Blank space : no data available.

TABLE 2. Interparticle force F_{pp}[a].

POWDER	Microtrac* F_{pp}	Sedigraph* 5100 F_{pp}	Malvern* 4700 F_{pp}	Galai* CIS-1 F_{pp}	Coulter* Multisizer F_{pp}
Flour (high protein)	1.42×10^{-10}			1.69×10^{-10}	
Diatomaceous earth (SiO_2)					
Kaolin	5.54×10^{-12}	1.38×10^{-12}		2.68×10^{-11}	
Calcium Carbonate ($CaCO_3$)	1.74×10^{-12}	3.46×10^{-13}		2.36×10^{-12}	4.79×10^{-13}
Sand			3.35×10^{-13}	1.60×10^{-12}	4.36×10^{-12}
Silicon Nitride (Si_3N_4)			4.67×10^{-14}		
Aluminum Oxide (α-Al_2O_3)		8.40×10^{-16}	6.72×10^{-15}		
Tungsten (W)					

[a]Interparticle force F_{pp} was calculated from H_o=3cm data because of incomplete results of H_o=5cm. (Unit of F_{pp} : N)

Blank space : no data available.

*The instruments used for the measurement of particle size.

TABLE 3. The boundary between Group C and Group A, $d_{C/A}$ (from $R_c = 1$).

POWDER	Microtrac* $d_{C/A}$	Sedigraph* 5100 $d_{C/A}$	Malvern* 4700 $d_{C/A}$	Galai* CIS-1 $d_{C/A}$	Coulter* Multisizer $d_{C/A}$
Flour (high protein)	33.53			35.51	
Diatomaceous earth (SiO_2)					
Kaolin	9.42	5.93		15.93	
Calcium Carbonate ($CaCO_3$)	6.30	3.68		6.98	4.10
Sand			3.63	6.11	8.54
Silicon Nitride (Si_3N_4)			1.81		
Aluminum Oxide (α-Al_2O_3)		0.44	0.88		
Tungsten (W)					

Unit of $d_{C/A}$: μm.

Blank space : no data available.

*The instruments used for the measurement of particle size.

TABLE 4. The boundary between Group C and Group A, $d_{C/A}$ (from $I_{C/A}$).

POWDER	Microtrac* $d_{C/A}$	Sedigraph* 5100 $d_{C/A}$	Malvern* 4700 $d_{C/A}$	Galai* CIS-1 $d_{C/A}$	Coulter* Multisizer $d_{C/A}$
Flour (high protein)	20.02			20.41	
Diatomaceous earth (SiO_2)					
Kaolin	11.56	9.91		13.78	
Calcium Carbonate ($CaCO_3$)	10.01	8.36		10.35	8.67
Sand			8.31	9.89	11.05
Silicon Nitride (Si_3N_4)			6.41		
Aluminum Oxide (α-Al_2O_3)		3.80	4.78		
Tungsten (W)					

Unit of $d_{C/A}$: μm.

Blank space : no data available.

*The instruments used for the measurement of particle size.

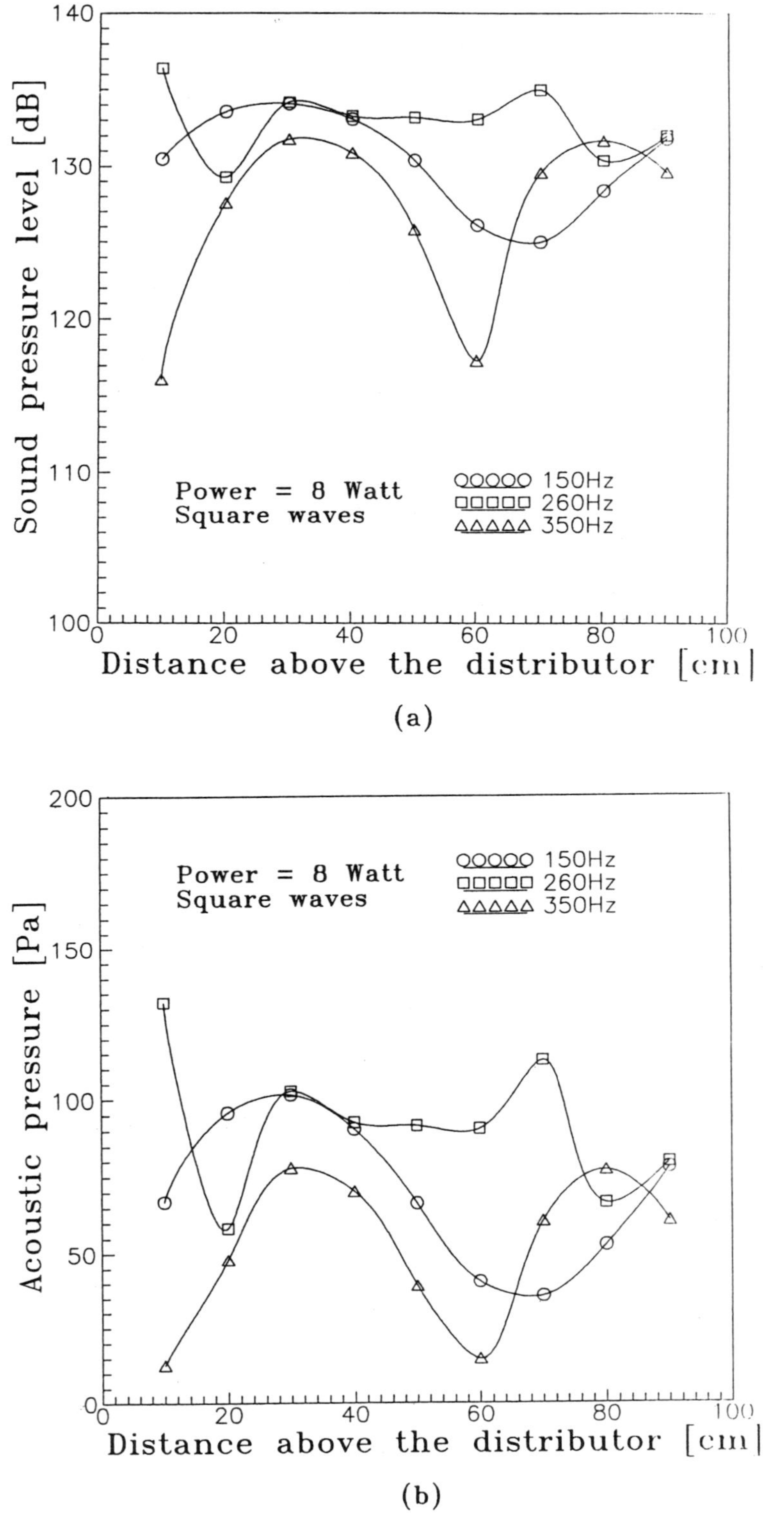

Figure 1. Intensity of sound above the distributor at some different frequencies.
(a) Distribution of SPL [dB]
(b) Distribution of acoustic pressure [Pa]

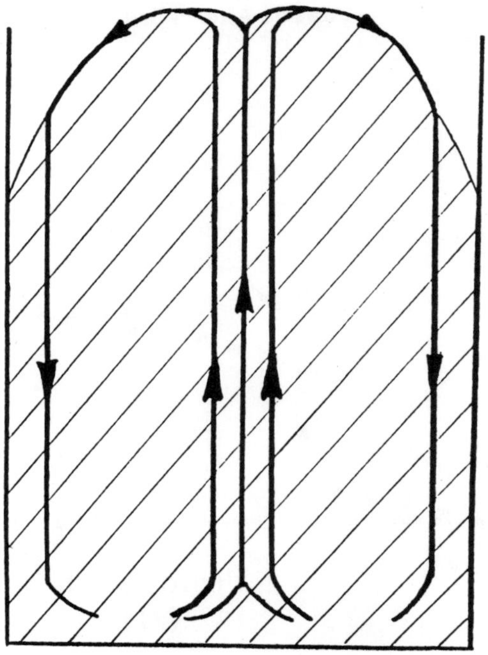

Figure 2. Spouted circulating pattern of flour in fluidized bed.

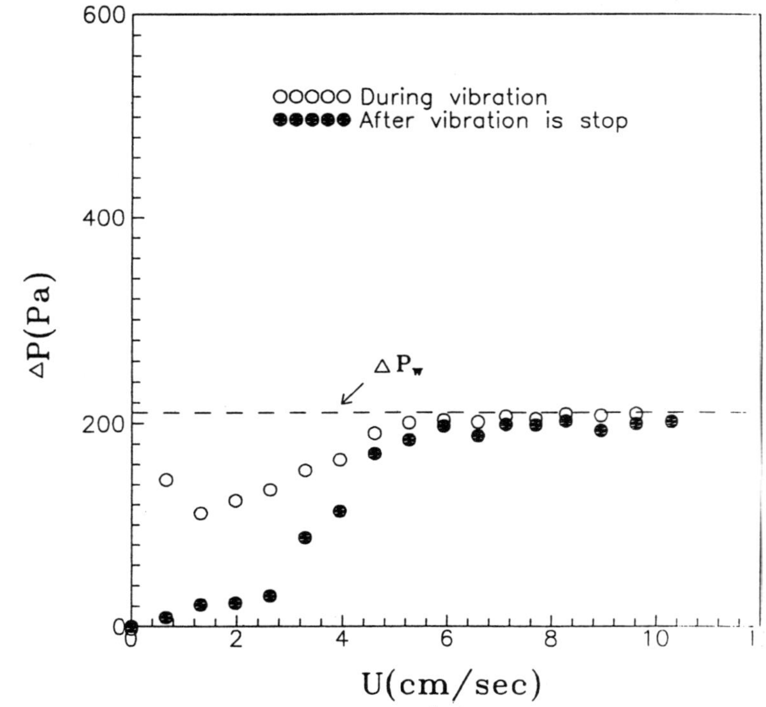

(a)

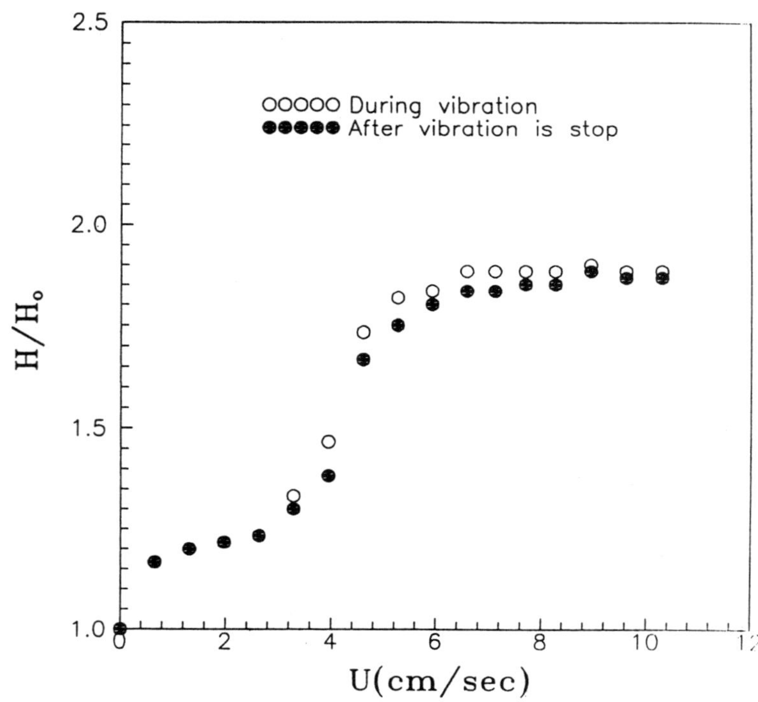

(b)

Figure 3. Comparison between "during vibration" and "after vibration is stopped" (powder : flour; $H_O = 3$ cm)
(a) Bed pressure drop versus superficial air velocity
(b) Bed expansion versus superficial air velocity

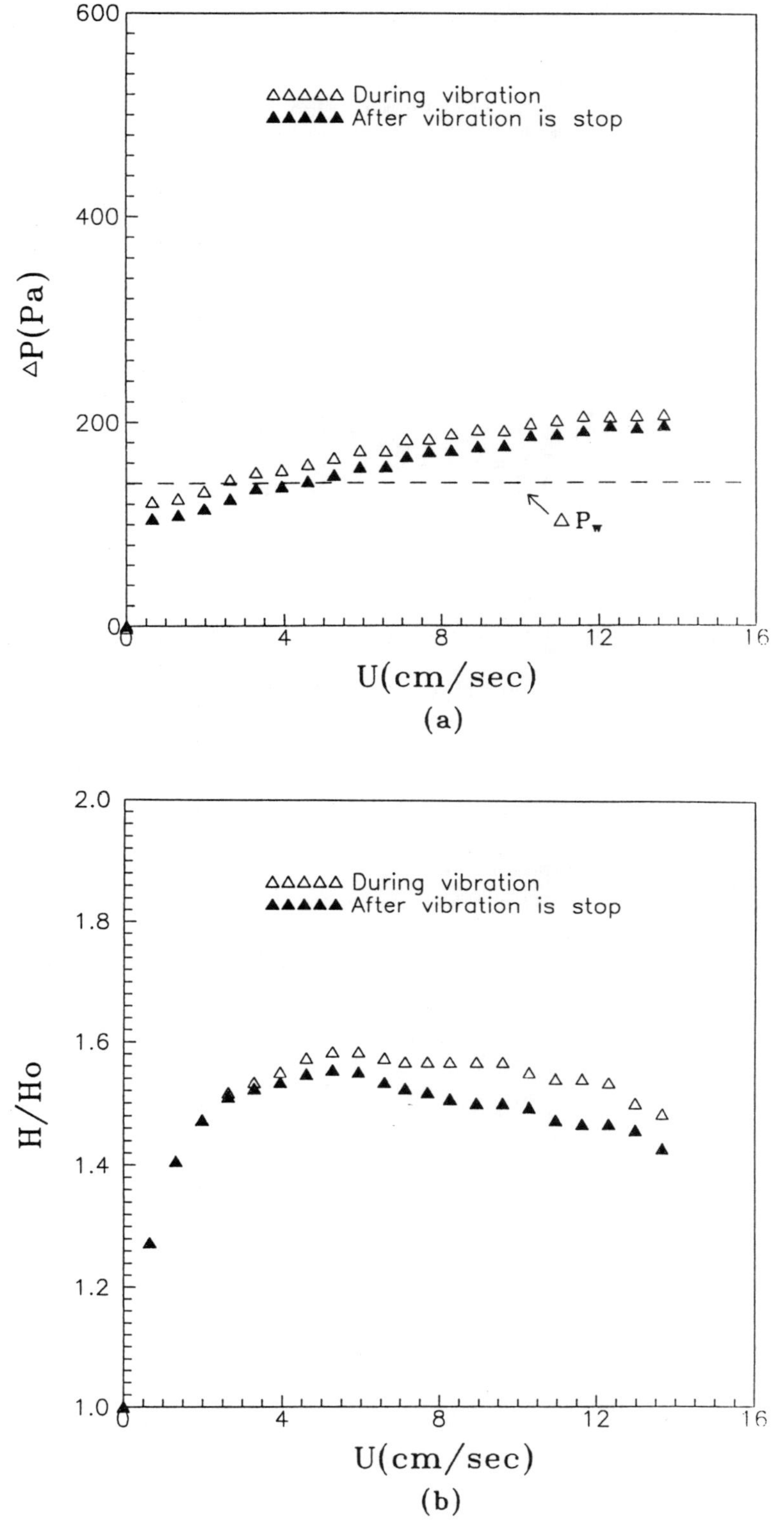

Figure 4. Comparison between "during vibration" and "after vibration is stopped" (powder : kaolin; H_o = 3 cm)
(a) Bed pressure drop versus superficial air velocity
(b) Bed expansion versus superficial air velocity

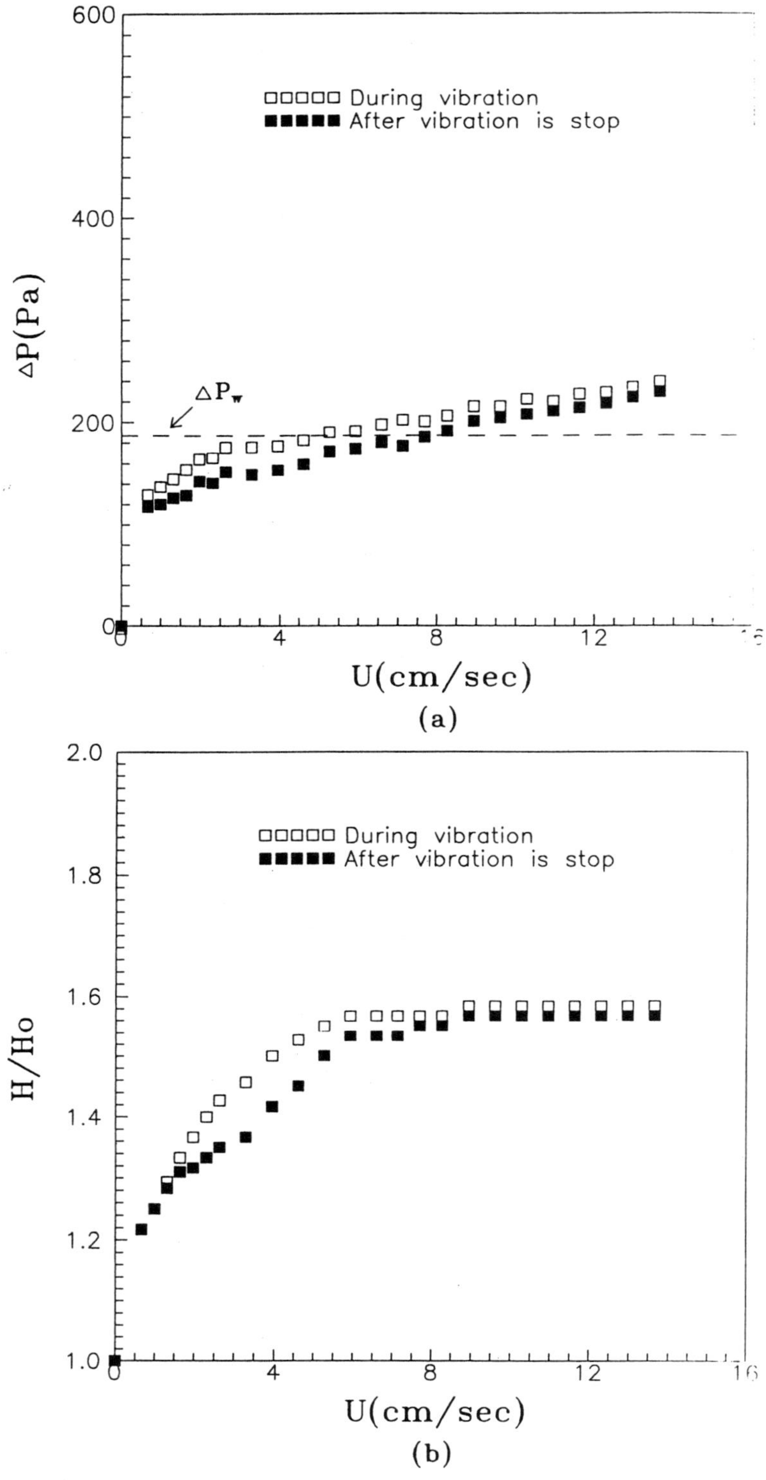

Figure 5. Comparison between "during vibration" and "after vibration is stopped" (powder : silicon nitride, $H_O = 3$ cm)
(a) Bed pressure drop versus superficial air velocity
(b) Bed expansion versus superficial air velocity

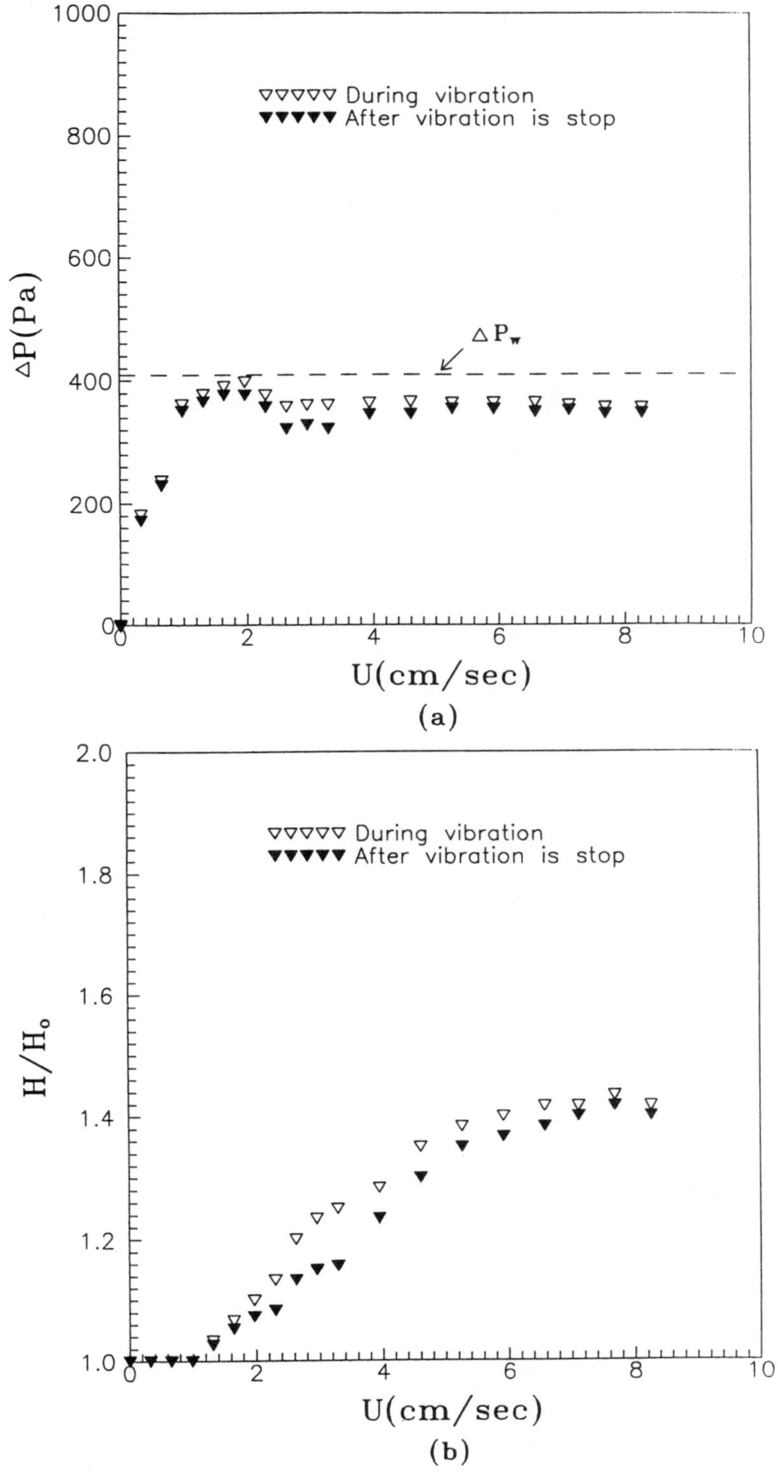

Figure 6. Comparison between "during vibration" and "after vibration is stopped" (powder : alumina oxide (α phase), H_0 = 3 cm)
(a) Bed pressure drop versus superficial air velocity
(b) Bed expansion versus superficial air velocity

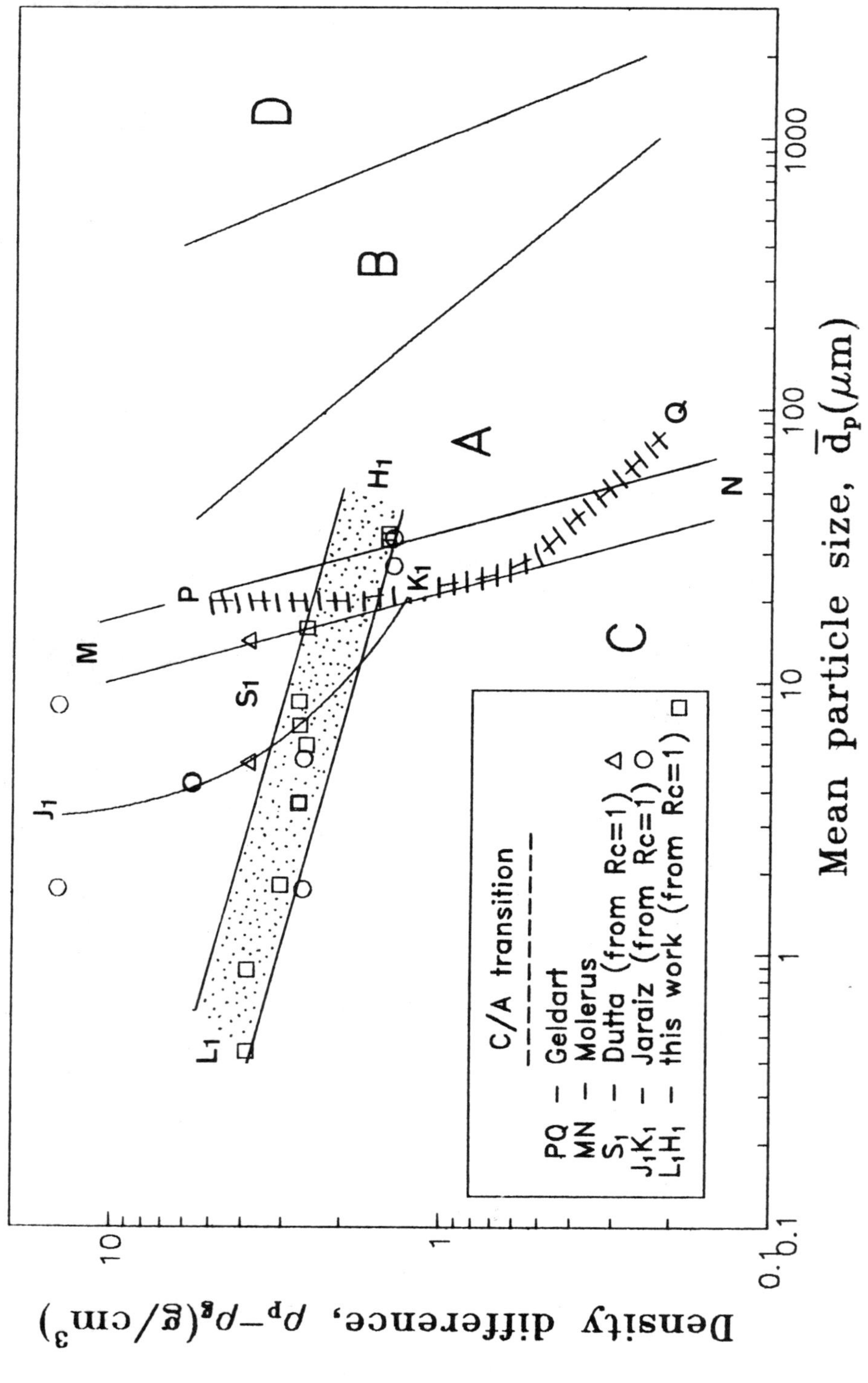

Figure 7. Geldart powder classification diagram based on $R_C = 1$

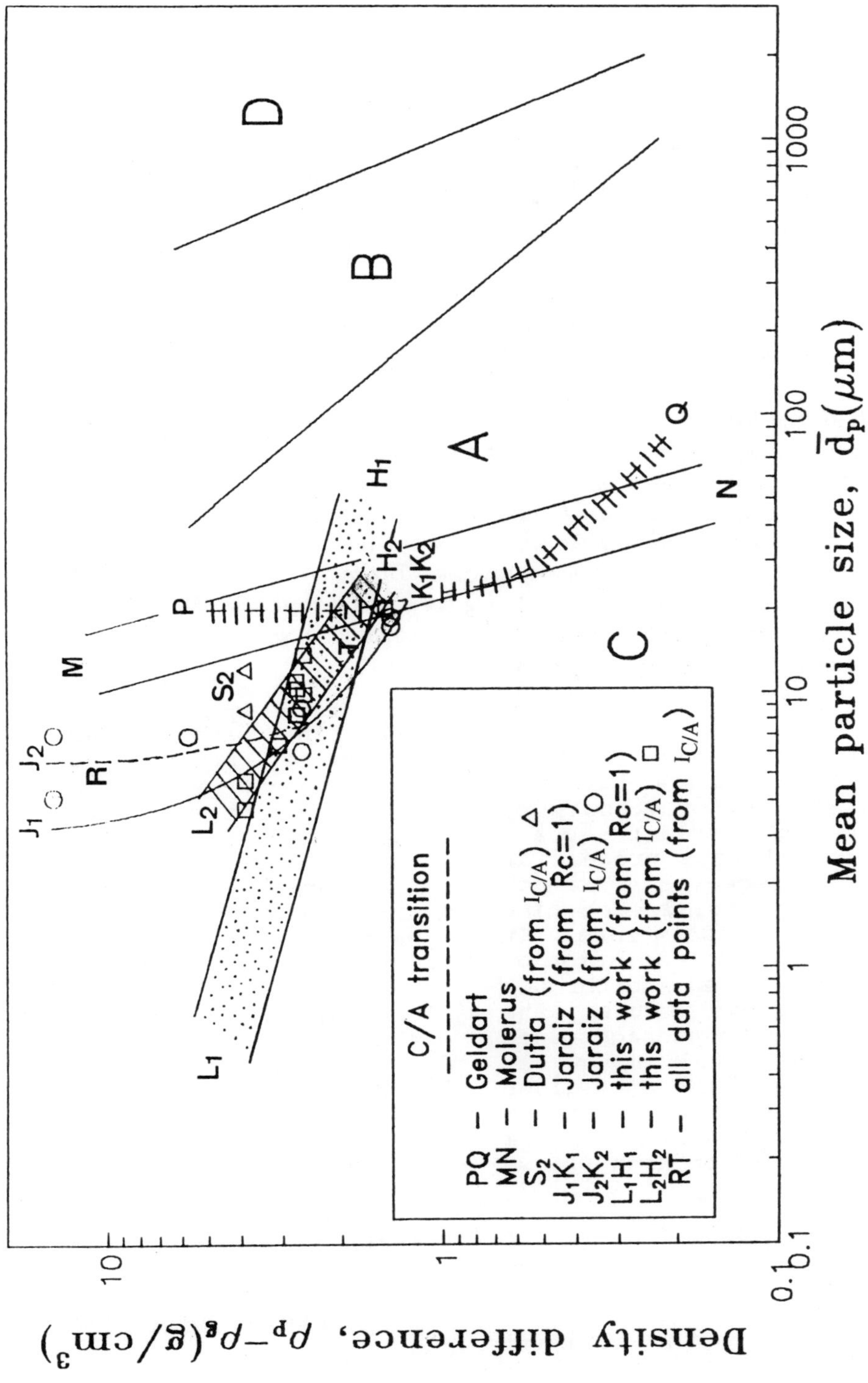

Figure 8. Geldart powder classification diagram based on $I_{C/A}$

The Uniformity of Particle Coating Occurring in Fluidized Beds

X.X.Cheng and R.Turton
Dept. of Chemical Engineering, P.O.Box 6101,West Virginia University, Morgantown, WV 26506

Research has been carried out to determine the uniformity of the deposition of coating material on 1mm diameter, approximately spherical particles coated in a 6" fluidized bed with a draft tube insert. The variation in the total amount of material deposited on individual particles during a batch coating operation was measured under a variety of fluidizing conditions. This variation was attributed to two sources, namely the variation in the number of times a particle passes through the spray zone and the variation in the amount of material that is deposited on a particle as it passes through the spray zone. The factors associated with these two components of variation were investigated. Finally, a phenomenalogical model was proposed to account for the variation in particle coating based on the hindering or sheltering effect exhibited by particles in the spray zone. Predictions of the model show fair agreement with the experimental data.

Theory

The coating of particles in a fluidized bed with a draft tube insert has been studied previously by Mann (1,2). He considered the coating to be a renewal type process in which a particle receives a random amount of coating material each time it passes through the spray zone. In addition, although the circulation of a particle around the bed and through the spray zone is quite regular, the time between successive passages does vary. Hence, the total amount of coating material deposited on a given particle is the result of a random number of additions of a random amount of material. From Mann (1) we may write:

$$\frac{\sigma_{tot}}{\mu_{tot}} = \left[\left(\frac{\sigma_n}{\mu_n}\right)^2 + \frac{1}{\mu_n}\left(\frac{\sigma_x}{\mu_x}\right)^2 \right]^{\frac{1}{2}} \qquad (1)$$

where σ and μ refer to the standard deviation and mean of the distribution, tot refers to the total amount of material deposited, n refers to the number of passages the particle takes through the spray zone and x refers to the amount of material deposited per pass through the spray.

From Equation (1) it can be seen that the total variation in the amount of material deposited on a particle during a batch coating operation is the sum of two terms. The first term on the right hand side of the equation represents the variation in coating due to the variation in the number of passages that different particles make through the spray zone. The second term on the right hand side of Equation (1) represents the variation due to the variation in the amount of material that is deposited each time a particle passes through the spray zone. The magnitude and relative importance of each of the terms on the right hand side of Equation (1) has not been addressed previously. It was the goal of this work to carry out measurements in order to evaluate the relative importance of these two effects.

EXPERIMENTAL APPARATUS AND PROCEDURE

In order to investigate the particle circulation characteristics and the variation in material deposited during the passage of particles through the spray zone several experimental techniques were employed. The apparatus and techniques used in this study are described below.

Fluidized Bed Coating Apparatus

A schematic diagram of the coating equipment is shown in Figure 1. Although a detailed description of the equipment is given elsewhere, by Cheng (3), a brief overview is included here. The fluidized bed consists of a section of 6" diameter plexiglass tubing with a draft tube located coaxially with the bed and a 9" expansion section for particle disengagement. The distributor plate is made of sintered stainless steel and the plenum is split into an inner and outer annular section. Each section is supplied by air through a separate rotameter, thus allowing independent control of the central and outer air streams. The coating liquid can be sprayed into the bed via an air atomizing nozzle placed at the center of the distributor plate. A water trap is placed at the exit of the bed to collect the excess ethanol which is used as the coating solvent.

Magnetic Tracer Technique

In order to measure the mean and variation in the number of particle circulations, a magnetic particle tracer technique was employed. The tracer particle was manufactured by encasing a small piece of a rare earth permanent magnetic in a ball of hardened flour and water dough. The movement of the tracer particle was detected by a coil of approximately 10,000 turns of 36-gauge insulated wire countersunk into the side of the aluminum draft tube. The detector coil was connected to a data acquisition system which recorded the induced voltage signal that occurred every time the particle passed through the draft tube. A diagram showing the details of the draft tube is given in Figure 2. A sample of typical data, after filtering, recorded during an experimental run is given in Figure 3.

Particle Coating Technique

In order to evaluate the relative variation in the amount of material received by a particle during batch-coating operations, it was decided to tag the coating material with an easy to detect dye.

This material (FD&C blue dye #1) gave a characteristic absorbance peak at 626nm and hence could be detected easily using conventional visible spectrophotometry. The procedure adopted to evaluate the relative variation of the amount of material deposited per particle during a coating operation was as follows:

(a) Coat a batch of approximately spherical 1mm sucrose particles (Nu-Pareil supplied by Ingredient Technology Co., Pennsauken, NJ) with a solution of ethanol containing 2 wt% of ethylcellulose and 0.15% of FD&C blue dye #1 for approximately 80 minutes.

(b) Take 60 random samples of 10 particles from the coated batch of material.

(c) Dissolve each sample in 4ml of spectrophotometric grade ethanol and allow to dissolve for 5 hours.

(d) Assay each dissolved sample at a wavelength of 626nm using a Perkin-Elmer Lambda 2 uv-visible spectrophotometer.

A histogram showing the variation in the amount of blue dye from the set of 60 samples was then constructed and an example of such a distribution is shown in Figure 4. From this data the mean and variance of the total dye deposited per particle was estimated from Equations (2) and (3):

$$\mu_{tot} = \frac{\mu_{samp}}{10} \qquad (2)$$

$$\sigma^2_{tot} = \frac{\sigma^2_{samp}}{\sqrt{10}} \qquad (3)$$

EXPERIMENTAL RESULTS

The data given in Figure 3, after suitable filtering and conditioning, as described by Cheng (3), can be used to construct the circulation time distribution (CTD) of the tracer particle. A typical CTD for a run is shown in Figure 5. From the CTD, it is necessary to evaluate the mean and variance of the distribution of the number of passages that different particles make around the bed during a batch coating process. The relationship between the means and variances of the CTD and the number distribution have been derived by Mann (2). For the situation where the average number of particle passages exceeds 20 (the typical average number of circulations for this work is approximately 1000) we may write:

$$\mu_N = \frac{T}{\mu_{CTD}} \qquad (4)$$

$$\sigma^2{}_N = \frac{T\,\sigma^2{}_{CTD}}{\mu^3{}_{CTD}} \qquad (5)$$

Thus, by knowing the characteristics of the CTD, we may infer the mean and variance of the number of passages distribution required in Equation (1).

The reproducibilty of the magnetic particle tracer technique and the effect of different tracer particle sizes on the circulation rate have been addressed by Cheng (3). In summary, this technique was found to be reproducible. In addition, by using a tracer particle with properties (diameter and density) close to those of the bed material circulation data which is representative of all particles in the bed is obtained.

Referring to Equation (1), it can be seen that the magnetic particle tracer technique enables the evaluation of the magnitude of the first term on the right hand side of the equation. In addition, the dye dissolution studies can be used to calculate the left hand side of Equation (1) and hence the magnitude of the second term on the right hand side of the equation can be deduced.

A direct evaluation of the variation in the amount of material deposited per particle per pass through the spray zone was not attempted in this work. Such a measurement is possible but would require very sensitive instrumentation in order to detect the minute quantities of material deposited in a single particle pass. In addition, many more (much greater than 60) particles would have to be sampled in order to get a reliable estimate of the variance (σ_x).

The results of five experimental runs in which circulation data, using the magnetic particle tracer, was taken simultaneously during coating operations are summarized in Table 1. The bottom line in Table 1 shows the relative magnitude of the first term on the right hand side of Equation (1) compared to the left hand side. The first term on the right hand side of the equation represents the relative variation associated with the circulation of particles and hence can be evaluated using the data from the tracer particle technique. From Table 1, it is evident that the variation in particle circulation rates accounts for less than 20% of the total variation. This leads us to conclude that it is the second term on the right hand side of Equation (1) that is responsible for the majority of the variation in the coating process.

PHENOMENALOGICAL MODEL OF THE COATING PROCESS

From the above results, it is clear that the major source of variation during the batch coating of particles can be attributed to the processes occurring when a particle passes through the spray zone. The variations occurring during the coating operation are attributed to the sheltering effect that particles close to the spray source have on particles further away from the spray nozzle. The basic picture of what is happening in the spray zone is illustrated in Figure 6.

For the model developed in this research, the following assumptions are made regarding the interaction of the particles and the coating material.

(1) The shape of the spray is a hollow cone.

(2) The particles are assumed to move through the spray at a constant velocity.

(3) The voidage in the spray zone is assumed to be constant.

(4) The particles are assumed to be distributed evenly through the spray from a distance r_o from the nozzle to the inside diameter of the draft tube insert R.

The above assumptions lead to a physical picture in which the solids pass through the spray zone as a uniformly distributed dispersion of constant voidage. With these assumptions and geometric arguments, it is possible to predict the relative variation of the mass of coating material that is deposited on single particles per pass through the spray zone. The details of this analysis are given by Cheng (3) and are summarized in the Appendix. The final result is shown below in Equation (6).

$$\frac{\sigma_x}{\mu_x} = \sqrt{\frac{9(1-\epsilon)^2(N^2-n_0^2)\exp(3(1-\epsilon)n_0)}{8[1-\exp(-1.5(1-\epsilon)(N-n_0))]^2}\int_{3(1-\epsilon)n_0}^{3(1-\epsilon)N}\frac{\exp(-t)}{t}dt-1} \tag{6}$$

Equation (6) gives, explicitly, the normalized standard deviation for the amount of material received per particle per pass through the spray zone. All the parameters in this equation are known for a given particle size and given system geometry except for the voidage in the spray zone. Although the voidage can be measured, this was not done in the current research. Consequently, the voidage was estimated using a simple one-dimensional model originally proposed by Yang (4) and Wen and Yu (5). Thus, the continuity and momentum equations given below were solved with the appropriate initial conditions in order to estimate the voidage in the spray zone.

Continuity

$$\frac{dz}{dt}(1-\epsilon)\rho_p A_{ins} = \frac{M}{\mu_t} \tag{7}$$

Equation of Motion

$$\frac{d^2z}{dt^2} = \frac{3}{4}C_{DS}\epsilon^{-4.7}\frac{\rho_g}{(\rho_p-\rho_g)d_p}(u_g-\frac{dz}{dt})^2-\frac{(\rho_p-\rho_g)}{\rho_p}g \tag{8}$$

By solving Equations (7) and (8) and substituting the calculated voidage into Equation (6), the normalized standard deviation of the amount of material that a single particle receives per pass through the spray zone can be estimated.

COMPARISON OF EXPERIMENTAL RESULTS WITH THE PREDICTIONS OF THE MODEL

The experimental data for five sets of batch-coating experiments are compared with the predictions of the phenomenalogical model in Figure 7. Also shown are the predictions of Equation (1) for the case when coating variation in the spray is ignored, i.e. for the case when σ_x is zero. It can be seen that the predictions of the model agree reasonably well with the data and predict the correct order of magnitude for the variation. It can also be seen that the second term on the right hand side of Equation (1) plays a major role in accounting for particle-to-particle variation and that the variation in circulation rate is considerably less important.

CONCLUSIONS

A magnetic tracer particle technique has been successfully developed and implemented to measure the circulation rate and its variation in fluidized bed coating equipment with draft tubes. Experimental data was taken to measure the variability of the amount of material deposited on single particles during batch-coating operations. This data indicated that the major source of variation was attributable to the variation in the amount of material deposited on single particles each time they passed through the spray. A simple phenomenalogical model was developed to explain and predict this variation based on the operating conditions in the fluidized bed. Predictions of this model compared reasonably well with experimental data. However, the model appears to be oversensitive to certain operating conditions.

The interactions between particles and an atomized liquid spray are complex. The model presented here to explain the variation in coating is simple in the sense that the voidage profile in the spray zone was assumed to be uniform across the cross-section of the draft tube. The actual voidage profile in the spray zone is almost certainly non-uniform and will be greatly affected by changes in the system geometry and fluidizing conditions. Currently, work is underway to measure the voidage profiles in the spray zone and to modify the phenomenalogical model to account for the non-uniformity of voidage in the spray zone.

ACKNOWLEDGMENTS

The authors would like to acknowledge the financial support of the National Science Foundation through grant CBT 8657548 and Schering-Plough Research Institute, Kenilworth, New Jersey. Donation of the F & C blue dye No. 1 from Warner-Jenkinson Co. is also gratefully acknowledged.

NOTATION

A_{ins} area of inside of draft tube, m^2
C_{ds} drag coefficient of a single particle
d_p diameter of particle, m
M mass of material in fluidized bed, kg
n_o dimensionless ratio given in Equation (6), $[r_o/d_p]$
N dimensionless ratio given in Equation (2), $[R/d_p]$
r_o radial distance from the center of the spray at which particles first start to move upward through the spray
R inside diameter of the insert, m
u_g gas velocity in draft tube, m/s

greek letters

ρ density, kg/m^3
ϵ voidage
μ mean of distribution
σ^2 variance of distribution
σ standard deviation of distribution

subscripts

CTD cycle time distribution
g gas
n the number of passes a particle makes through the spray zone (or around the fluidized bed)
p particle
samp sample
tot refers to the total mass of material deposited on a single particle
x mass of coating material

Bibliography

1. Mann, U., Coating of Particulate Solids by Air Suspension, PhD Thesis, University of Wisconsin, 1972
2. Mann, U., Analysis of Spouted-Bed Industrial and Eng. Chem. Process Design and Dev., 22, 288-292, 1983
3. Cheng, X.X., The Studies of Uniformity of Particle Coating in Fluidized Beds, Ph.D. Dissertation, West Virginia University, 1993
4. Yang, W.C.,A Unified Theory on Dilute Phase Pneumatic Transport, Journal of Powder and Bulk Solids Technology, 1, 89-95, 1977
5. Wen C.Y. and Y.H. Yu, CEP Symp. Ser., 62(62), 100-111, 1966

APPENDIX

A diagrammatic sketch of the system geometry and the notation used in the derivation are shown in Figure 8. Consider the shaded element shown in Figure 8.

volume of element

$$= 2\,\pi\,\alpha\,\beta\,r^2\,\Delta\,r$$

number of particles in element

$$= \frac{12\,\alpha\,\beta\,r^2\,(1-\epsilon)\,\Delta r}{d_p^3}$$

projected surface area of particles in element

$$= \frac{3\,\pi\,\alpha\,\beta\,(1-\epsilon)\,r^2\,\Delta r}{d_p}$$

equivalent projected area of particles at r_o

$$= \frac{3\,\pi\,\alpha\,\beta\,(1-\epsilon)\,r_o^2\Delta r}{d_p}$$

fraction of spray with which particles intersect

$$= \frac{3\,\pi\,\alpha\,\beta\,(1-\epsilon)\,r_o^2\,\Delta r}{2\,\pi\,\alpha\,\beta\,r_o^2\,d_p}$$

$$= K_1\,\Delta r$$

$$\text{where } K_1 = \frac{3\,(1-\epsilon)}{2\,d_p}$$

let the fraction of the spray that hits particles at r be ΔF_r where:

$$\Delta F_r = K_1 \Delta r\,(1 - K_1 \Delta r)^{\frac{r-r_o}{\Delta r} - 1}$$

dividing by Δr and taking the limit as $\Delta r \rightarrow 0$ we get:

$$\frac{dF}{dr} = K_1 e^{-(r-r_o)K_1}$$

now at any radius r the amount of material received per particle per pass is given by x where:

$$x = \frac{M_{spray} t_{spray}}{\Delta n}\left[\frac{dF}{dr}\right]\Delta r$$

$$= \frac{M_{spray}\,t_{spray}\,d_p^2\,e^{-(r-r_o)K_1}}{8\,\alpha\,\beta\,r^2}$$

where M_{spray} is the mass flow rate of spray and t_{spray} is the time that the particle spends in the spray.

from the geometry of the system we may write:

$$t_{spray} = \frac{\alpha\,\beta\,r}{u_p}$$

where u_p is the particle velocity through the spray.

substituting t_{spray} in x and simplifying we get:

$$x = \frac{M_{spray}\,d_p^2\,e^{-(r-r_o)k_1}}{8\,u_p r}$$

It is further assumed that the particles are uniformly distributed throughout the cross-section of the insert from r_o to R. Thus, the probability of finding a particle at a given radius is P(r)dr where:

$$P(r) = \frac{2\,r}{R^2 - r_o^2}$$

The mean and variance of the amount of material deposited on a single particle per pass through the spray zone may be calculated as follows:

$$\mu_x = \int_{r_o}^{R} x P(r)\,dr = \int_{r_o}^{R} \frac{M_{spray} d_p^2 e^{-(r-r_o)K_1}}{8\,u_p r}\,\frac{2\,r}{R^2 - r_o^2}\,dr$$

$$= \frac{M_{spray} d_p^2\,(1 - e^{-(R-r_o)K_1})}{4\,u_p\,(R^2 - r_o^2)\,K_1}$$

and

$$\sigma_x^2 = \int_{r_o}^{R} \left[\frac{M_{spray} d_p^2 e^{-(r-r_o)K_1}}{8\,u_p\;\;r} - \mu_x\right]^2 \frac{2\,r}{R^2 - r_o^2}\,dr$$

$$= \frac{2}{R^2 - r_o^2}\left[\frac{M_{spray} d_p^2 e^{r_o K_1}}{8\,u_p}\right]^2 \int_{2 r_o K_1}^{2RK_1} \frac{e^{-t}}{t}\,dt - \mu_x^2$$

combining the two equations above we get Equation (6):

$$\frac{\sigma_x}{\mu_x} = \sqrt{\frac{9\,(1-\epsilon)^2\,(N^2 - n_0^2)\,\exp(3\,(1-\epsilon)\,n_0)}{8\,[1 - \exp(-1.5\,(1-\epsilon)\,(N - n_0))]^2}\int_{3(1-\epsilon)n_0}^{3(1-\epsilon)N} \frac{\exp(-t)}{t}\,dt - 1} \tag{6}$$

Table 1 : Results of Batch Coating Experiments

Run No.	1	2	3	4	5
σ_{tot}/μ_{tot} *	0.2160	0.2850	0.2900	0.2070	0.2990
σ_N/μ_N **	0.0178	0.0193	0.0386	0.0400	0.0374
$\frac{\sigma_N/\mu_N}{\sigma_{tot}/\mu_{tot}}$	8.2 %	6.8 %	13.3 %	19.3 %	12.5 %

* Relative variation of the amount of coating material deposited on particles calculated from the dissolution of blue dye.

** Relative variation associated with the circulation of particles in a fluidized bed coater calculated from magnetic tracer particle experiments.

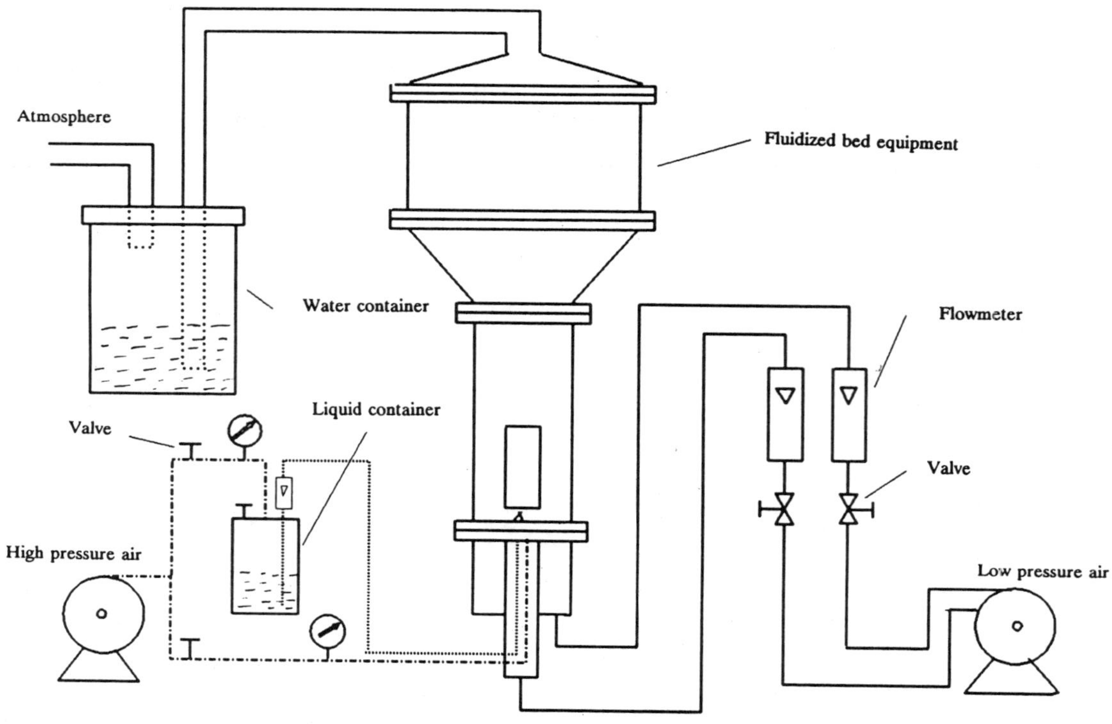

Figure 1 : Schematic Diagram of Fluidized Bed Coating Equipment

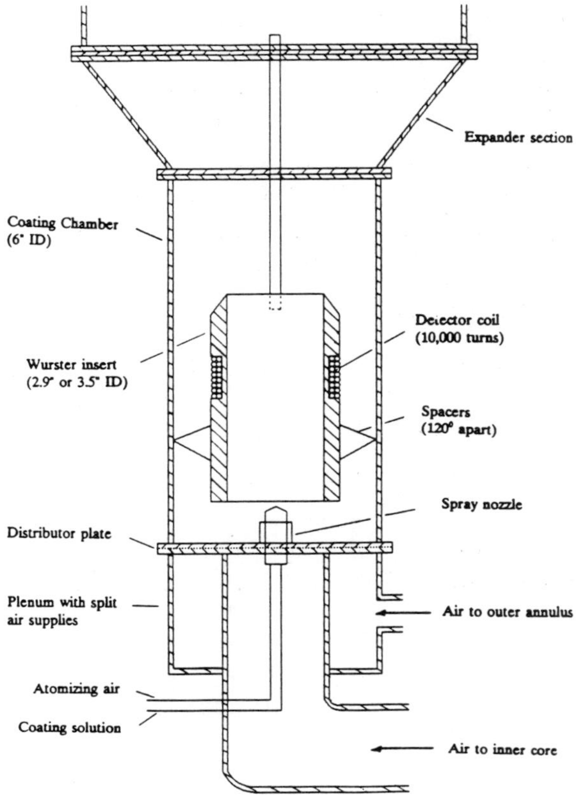

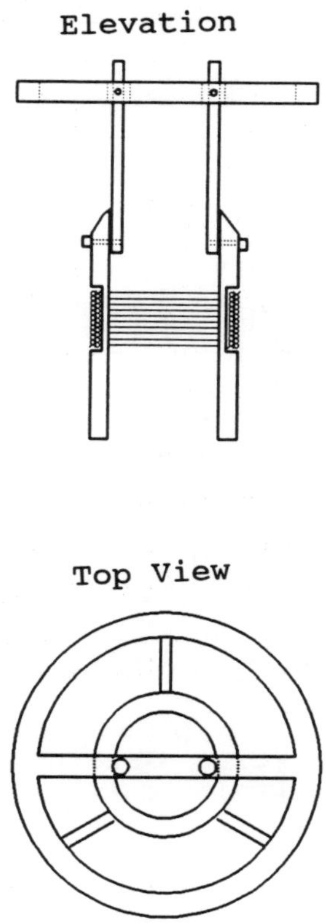

Figure 2 : Diagram Showing Details of Fluidized Bed Internals

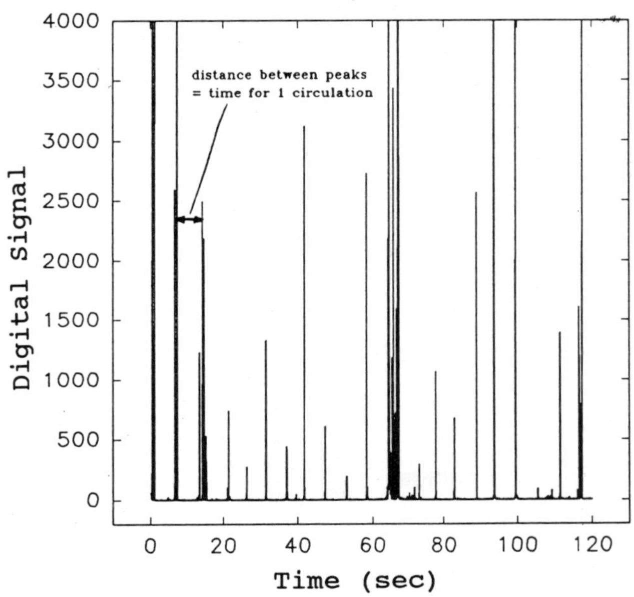

Figure 3 : A Typical Set of Data Obtained from the Magnetic Tracer Particle Detector During a Fluidization Experiment

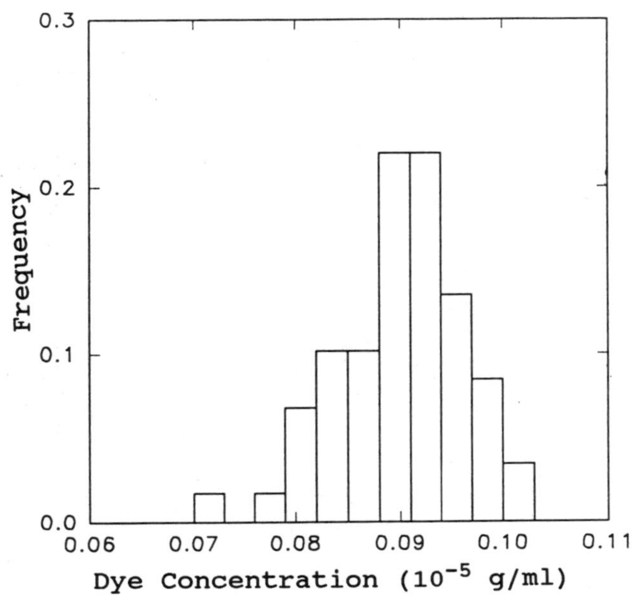

Figure 4 : Histogram Showing the Variation in Blue Dye Content of the 60 Samples of 10 Particles Taken from a Batch of Coated Particles

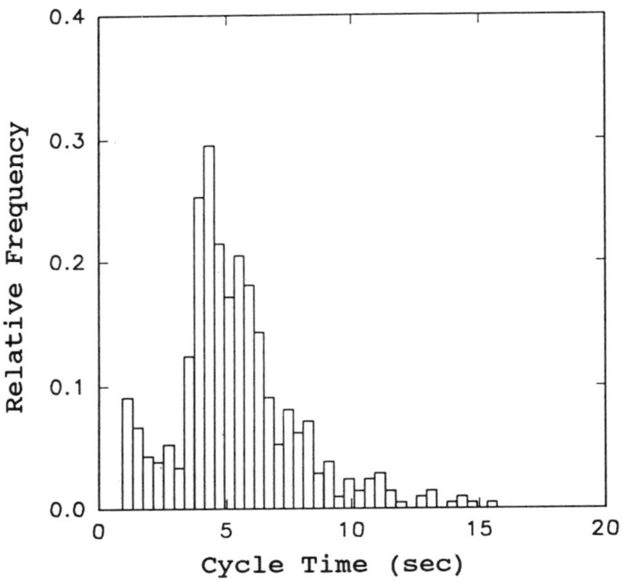

Figure 5 : A Typical Cycle Time Distribution (CTD) Obtained for Particles in a Batch Coating Operation

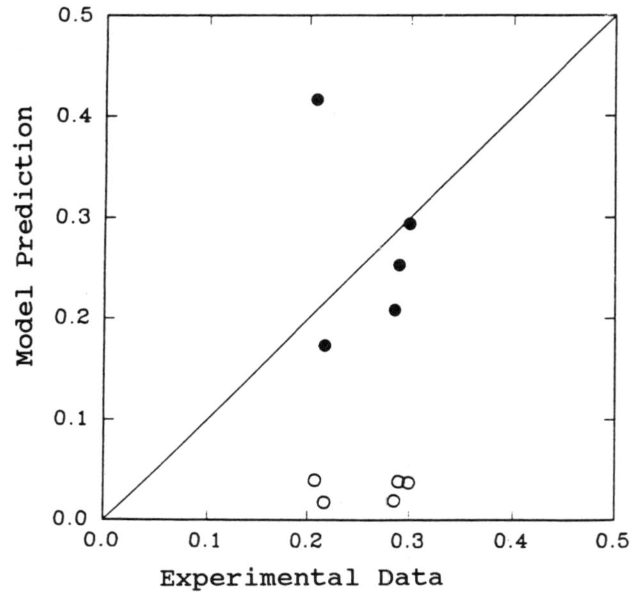

o relative variation predicted using CTD only
● relative variation predicted by model

Figure 7 : Comparison of Experimental Data with the Predictions of Equation (1) and Phenomenalogical Model.

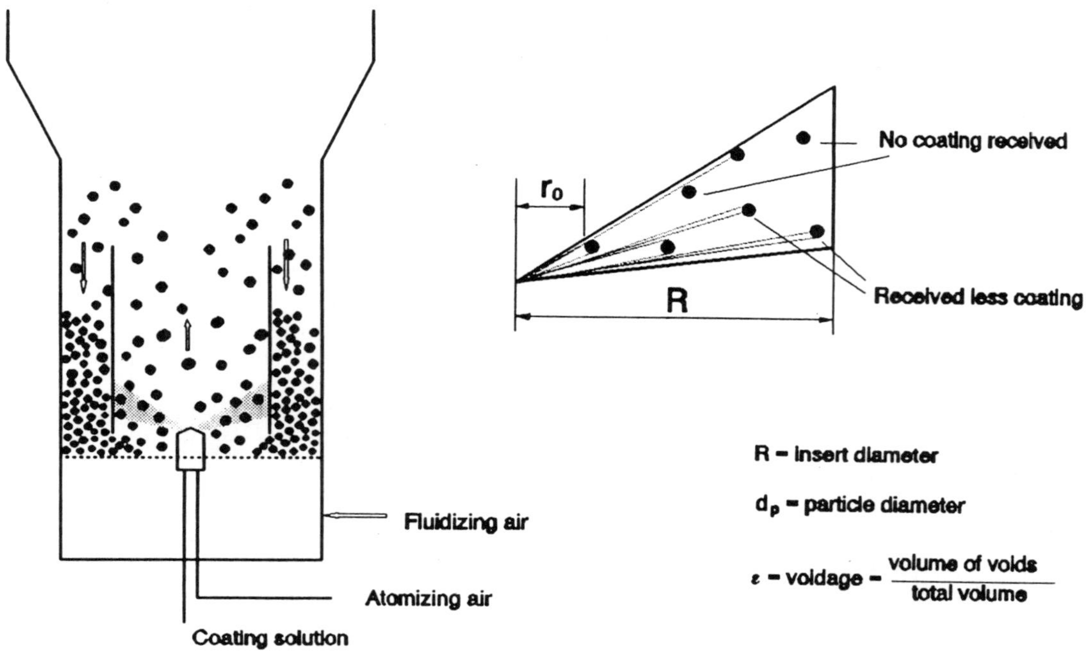

Figure 6 : Diagram Illustrating the Sheltering Effect of Particles Close to the Center of the Spray on Other Particles in the Spray Zone.

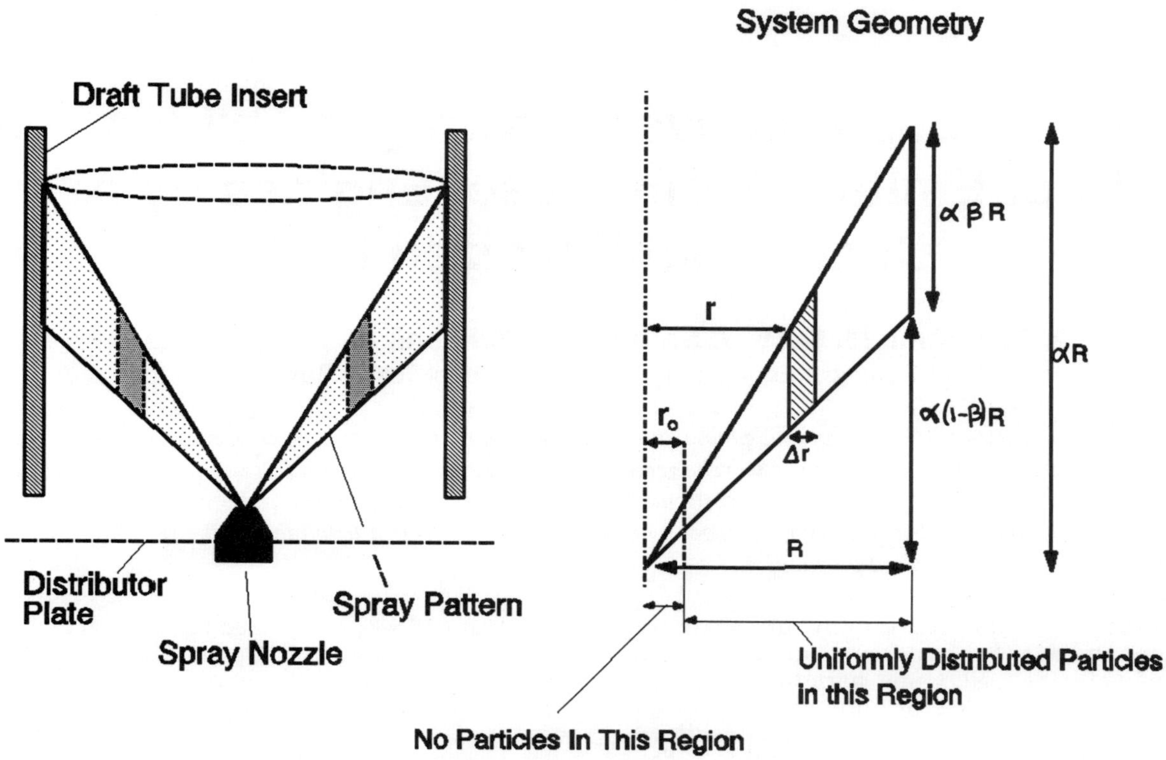

Figure 8 : System Geometry and Notation Used in the Derivation of the Coating Model given in the Appendix.

Simultaneous Removal of Nitrogen Oxides and Soot Emitted From Diesel Engines by a Centrifugal Fluidized Bed

A. Tsutsumi, M. Sato, Y. Shibuya, M. Kikuchi, and K. Yoshida
Department of Chemical Engineering, The University of Tokyo, Tokyo 113, Japan

Simultaneous removal of nitrogen oxides and soot emitted from diesel engines have been attempted using a centrifugal fluidized bed with a copper ion-exchanged ZSM-5 zeolite catalyst as fluidized solids. The NO reduction and the soot removal efficiency were investigated using actual exhaust gas from a diesel engine. A significant reduction of NO was observed with high soot removal efficiency. This is attributable to the reaction between NO and soot trapped on the bed particles.

Air pollution in urban areas is an increasingly serious problem due to the enormous increase in the number of vehicles. NOx and harmful pollutants such as soot and soluble organic fraction from diesel-powered vehicles are especially considered to be one of the major causes of air pollution. The particulate and NO_x emissions from diesel engines are higher than those from gasoline engines, but diesel engines have higher thermal efficiency and emit less carbon monoxide and unburned hydrocarbons. There has been, therefore, much effort made toward development of improved emission control in diesel engines through combustion modifications and other devices. However, many controls which reduce NO cause an increase in the concentration of incomplete combustion products such as hydrocarbons, carbon monoxide, and soot because NO is a product of an efficient combustion process. Furthermore, most NO emission controls reduce the fuel economy, which is a major attraction of diesel engines. Thus, treatment techniques of the exhaust gas from diesel engines are of increasing importance.

The removal of NO from the exhaust streams of various combustion sources is achieved by the selective catalytic reduction of NO with ammonia or hydrocarbons. In recent literature, the high catalytic activity of Cu ion-exchanged ZSM–5 zeolite for removing NO has been reported by direct decomposition of NO [1, 2] or by selective reduction with hydrocarbons [3-5] even in the presence of oxygen.

Trap systems have been examined to remove soot from diesel exhausts, and fluidized bed applications have been reported by several investigators [6-8]. Fluidized beds can achieve high collection efficiency for micron and submicron size particles from gases. Pfeffer et al. [9] demonstrated the feasibility of using a centrifugal fluidized bed composed of inert granules as a high efficiency filter to separate aerosol from a contaminated gas stream. In a centrifugal fluidized bed high gas throughputs can be obtained by increasing the rotational speed. The space requirement of a centrifugal fluidized bed is much smaller than that of a columnar fluidized bed. In addition, the large centrifugal force produced by the rotation reduces bypassing of gas due to bubble formation.

In this paper, we propose the idea of using a centrifugal fluidized bed as a treatment system for diesel exhausts. The centrifugal fluidized bed is considered to be capable of removing soot while simultaneously reducing NO_x. The operational flexibility and the compactness are also suited for diesel-powered engines, which are widely used in personal cars and modular power stations. The NO reduction and the collection efficiency of soot in the effluent from a diesel engine were examined in a centrifugal fluidized bed with a copper ion-exchanged ZSM-5 zeolite catalyst.

EXPERIMENTAL

Apparatus and procedure

A schematic diagram of the experimental apparatus is shown in Fig. 1. The centrifugal fluidized bed consisted of a plenum chamber, a rotor, and a gas distributor. Both the plenum chamber and the rotor were made of stainless steel. The rotor was of 184 mm internal diameter and

50 mm wide. The gas distributor was a cylindrical sintered stainless plate of 150 mesh pore size. Ports were provided for inlet and outlet gas sampling. A known weight of catalyst particles was loaded into the bed, and the bed was rotated in the vertical plane. After the rotational speed was established, air was fed into the bed to fluidize the particles. In the NOx reduction experiment, actual exhaust gas from a diesel engine was used. All experiments were performed with 0.2 or 0.3 kg of Cu-ZSM-5 zeolite, and the superficial gas velocity was varied in the range of 0.02-1.5 m/s. The speed of the rotor was controlled with an invertor and measured by a stroboscope. The angular velocity was ranged from 62.8 to 104.7 rad s^{-1}.

The pressure drop was measured with a manometer. The bed expansion was determined by a photographic method, replacing the side walls of both the rotor and the plenum chamber by transparent acrylic resin plates. In NO_x reduction experiments, the concentration of NO was measured by a NO_x meter (ECL-77A, Yanagimoto Co.). The exhaust gas from a diesel engine (Yanmar Diesel Co.) was heated to the desired temperature with a heater and then introduced into the centrifugal fluidized bed reactor. The exhaust of the diesel engine containing 530 ppm of NO, 10 ppm of NO_2, and 10% of oxygen was fed to decompose NO at 692 K.

The gas at the inlet and outlet was withdrawn and filtered by a glass fiber filter. The soot removal efficiency was evaluated by thermogravimetric analysis. By heating samples up to 873 K for an hour the soot concentration was determined from the weight loss. The soot removal efficiency is defined as

$$\eta = 1 - \frac{C_{out}}{C_{in}} \qquad (1)$$

where C_{in} and C_{out} are the concentrations of soot particles in a unit volume of gas before and after passing through the centrifugal fluidized bed, respectively. The entrainment of fine catalyst particles was determined from the weight difference between the initial and heated filter which filtered the gas at the outlet of the bed.

Catalyst particles were withdrawn and the gas distributor was detached from the centrifugal fluidized bed after passing the gas through the bed for a predefined time. The catalyst particles and distributor were weighed, and then regenerated by burning off the soot at 873 K for an hour. They were reweighed and the amounts of soot trapped on the catalyst particles and distributor during a run were obtained from the weight loss. These experiments were carried out at 693 K with a heating period of 45 min.

Preparation of the catalyst

The ZSM-5 zeolite was synthesized by hydrothermal reaction. After filtration, 6 kg of ZSM-5 zeolite particles were suspended in water and ion-exchanged with the addition of 1 kg of Copper (II) acetate for 24 h at room temperature. They were then washed with distilled water and dried at 383 K for 12 h. Table 1 shows the composition of the zeolite. The catalyst was prepared by spray granulation of Cu ion-exchanged ZSM-5 zeolite using 20 kg colloidal silica (20wt% of SiO_2) as a binder at 523 K and calcined at 773 K for 4 h. Table 2 summarizes the physical properties of the catalyst.

RESULTS AND DISCUSSION

Figure 2 shows the pressure drop plotted against the gas velocity at different angular velocities. The pressure drop rises to a peak at U_{mfc} and then decreases slightly as gas velocity continues to increase above U_{mfc}. The increase in angular velocity increases the pressure drop because of the radial acceleration [10]. Observed minimum fluidization velocities were in good agreement with values calculated from the correlation given by Levy et al. [10].

The variation of bed height with gas velocity and angular velocity is shown in Fig. 3. Above U_{mfc} the bed expands as the gas velocity increases and the angular velocity decreases. Figure 4 shows the relationship between bed expansion ratio and dimensionless gas velocity. The bed expansion ratio [11] is defined as

$$N_v = \frac{V}{V_0} \qquad (2)$$

where V and V_0 represent the bed volumes at U_g and U_{mfc}, respectively.

The experimental results of the reduction of NO in the diesel engine exhaust at 693 K are shown in Fig. 5. A significant reduction in NO was observed at low gas velocities. These values are considerably higher than the NO reduction by hydrocarbons over Cu-ZSM-5 catalyst reported in the literature [3-5]. The conversion of NO decreased with the increase in gas velocity. No significant effect of angular velocity on the degree of NO reduction was observed. This means the expansion of bed height has no significant effect on NO reduction. Figure 6 shows the effect of temperature on NO conversion. As the temperature decreases, the NO conversion decreases significantly. Figure 7 shows the effect of bed mass on the NO conversion by plotting NO conversion versus gas hourly space velocity based on catalyst volume at fixed bed (GHSV). No significant effect due to bed mass was observed. This suggests that the reduction of NO takes place homogeneously in the bed.

Carry-over of catalyst particles from the bed was very small during experiments. The mass of catalyst particles in the bed remained constant at 162±5 g during experiments when the initial bed mass was 200 g. It can be assumed that the entrainment of fine particles takes place in the initial stage of fluidization.

The effect of gas velocity on soot removal efficiency in the centrifugal fluidized bed is shown in Fig. 8. It can be seen that the centrifugal fluidized bed reactor can separate soot with a very high removal efficiency (99%) at a superficial gas velocity of 0.184 m s^{-1}. At higher gas velocities, soot removal efficiency is reduced to 77-83%.

Figure 9 shows the mass fractions of soot as a function of time. It was found that a negligible amount of soot was collected on the gas distributor. Both the emission of soot from the bed and the mass of deposited soot on the gas distributor increase with time. As can be seen in Fig. 9, half of the soot throughput is trapped on the catalyst particles at 90 min. The results show no soot accumulation in the bed; on the contrary, the mass of soot trapped on catalyst particles decreases with time until it reaches 0 at 200 min. This suggests that the elimination of soot from catalyst particles due to the reaction with NOx takes place in the bed. Thermodynamically, the NOx-carbon reaction is favorable. Several investigators have reported that carbon has a certain degree of selectivity for reducing NO emissions in the presence of excess oxygen [12-16]. Yamashita et al. [13] observed that the reaction between NO and metal-loaded carbon was remarkably promoted by the presence of oxygen at a temperature as low as 573 K. Chu and Schmidt [16] found NOx reactions with graphite had much higher intrinsic rates than those with O_2. Thus, it is likely that soot trapped on the catalyst particles acts as a reducing agent for NO. During the heating period soot was deposited on the catalyst particles in the bed. As the temperature rose, the soot trapped on the catalyst particles began to react with NO and the amount of soot in the bed decreased. This can explain the NO conversion considerably higher than those reported in the literature. In addition, the fact that the color of catalyst particles withdrawn from the bed changed to white with time confirms that the deposited soot was eliminated from catalyst particles due to the reaction with NOx.

CONCLUSIONS

The centrifugal fluidized bed system was applied to the simultaneous removal of nitrogen oxides and soot emitted from diesel engines. The experiments for NO reduction and soot removal in actual exhaust from a diesel engine were carried out in the centrifugal fluidized bed with Cu ion-exchanged ZSM-5 zeolite catalyst as the fluidized solids. The NO reduction and the collection efficiency of soot in the effluent from a diesel engine were investigated. Hydrodynamic characteristics were also examined as a function of gas velocity, bed mass and angular velocity.

A high NO conversion efficiency was achieved at low gas velocity. In addition, high soot removal efficiency was obtained. From the mass balance experiment, no soot accumulation in the bed was observed. These facts suggested that soot trapped on the bed particles reduces NO. Results of the present study show that the centrifugal fluidized bed has the potential for removing soot simultaneously with the reduction of NO_x emitted from diesel engines.

ACKNOWLEDGMENT

The work was supported by Grant-in-Aid for Encouragement of Young Scientists of the Ministry of Education, Science and Culture, Japan (No. 04750787).

NOTATION

C_{in}	soot concentration in gas at inlet	kg m^{-3}
C_{out}	soot concentration in gas passing through the bed	kg m^{-3}
N_v	bed expansion ratio	-
U_g	superficial gas velocity	m s^{-1}
U_{mfc}	minimum fluidization velocity	m s^{-1}
V	bed volume	m^3
V_0	bed volume at U_{mfc}	m^3
η	soot removal efficiency	-
ω	angular velocity	rad s^{-1}

LITERATURE CITED

1. Iwamoto, M., H. Yahiro, Y. Mine and S. Kagawa, *Chem. Letter*, 213 (1989).

2. Iwamoto, M., H. Yahiro, K. Tanda, N. Mizuno, Y. Mine and S. Kagawa, *J. Phys. Chem.*, **95**, 3727 (1991).

3. Held, W., A. Konig, T. Richter and L. Puppe, *SAE Paper* 900496 (1990).

4. Sato, S., Y. Yu-u, H. Yahiro, N. Mizuno and M. Iwamoto, *Applied Catalysis*, **70**, L1 (1991).

5. Cho, B. K., *J. Catal.*, **142**, 418 (1993).

6. Cole, W. E. and M. Suo, *Chem. Eng. Prog.*, December, 38 (1979).

7. Carlomagno, G. M., R. Festa and L. Massimilla, *AIChE Symp. Ser.*, No. 222, **79**, 66 (1983).

8. Massimilla, L., S. Vaccaro and A. Ciajolo, *Transport in Fluidized Particle Systems*, Edited by L. K. Doraiswamy and A. S. Mujumdar, Elsevier Science, Amsterdam, p. 487 (1989).

9. Pfeffer, R., G. I. Tardos and E. Gal, *Proc. Fifth Eng. Found. Conf. on Fluidization*, Elsinore, Denmark, p.667 (1986).

10. Levy, E., N. Martin and J. Chen, *Fluidization*, Edited by J. F. Davidson and D. L. Keairns, Cambridge University Press, London, p.71 (1978).

11. Takahashi, T., Z. Tanaka, A. Itoshima and L. T. Fan, *J. Chem. Eng. Jpn.*, **17**, 333 (1984).

12. Inui, T., T. Otowa and Y. Takegami, *Ind. Eng. Chem. Prod. Res. Dev.*, **21**, 56 (1982).

13. Yamashita, H., H. Yamada and A. Tomita, *Appl. Catal.*, **78**, L1 (1991).

14. Teng, H., E. M. Suuberg and J. M. Calo, *Energy & Fuels*, **6**, 398 (1992).

15. Illán-Gómez, M. J., A. Linares-Solano and S.-M. de Lecea, *Energy & Fuels*, **7**, 146 (1993).

16. Chu, X. and L. D. Schmidt,, *Ind. Eng. Chem. Res.*, **32**, 1359 (1993).

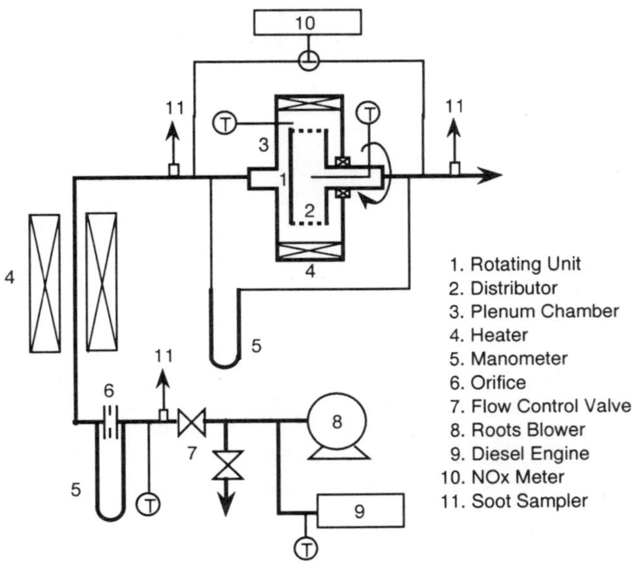

FIGURE 1. Schematic diagram of the experimental apparatus.

Table 1. Composition of Cu-ZSM-5 zeolite

SiO_2 [wt%]	Al_2O_3 [wt%]	Na_2O [wt%]	CuO [wt%]	ignition loss [wt%]	exchange level [wt%]
84.7	4.9	1.3	3.2	5.6	65.8

Table 2. Physical properties of the catalyst

mean particle size[μm]	density [kg m^{-3}]	void fraction [-]
26	1900	0.56

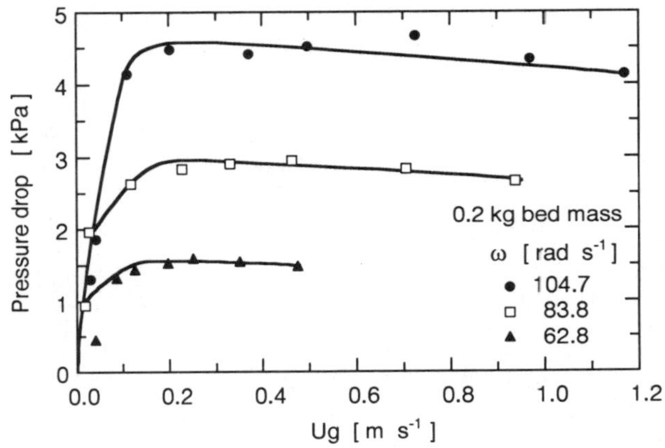

FIGURE 2. Pressure drop versus gas velocity.

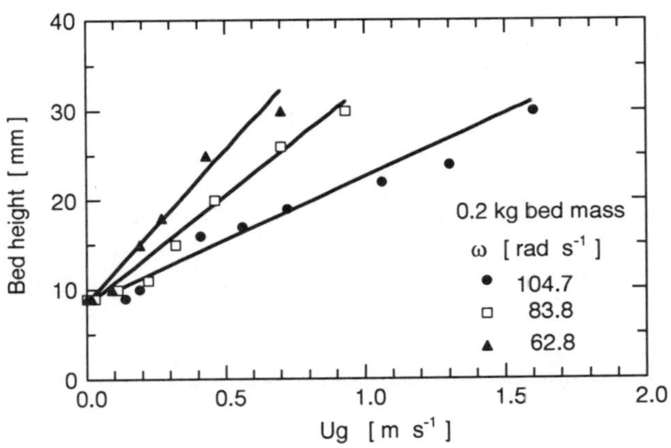

FIGURE 3. Variation of bed hight with gas velocity.

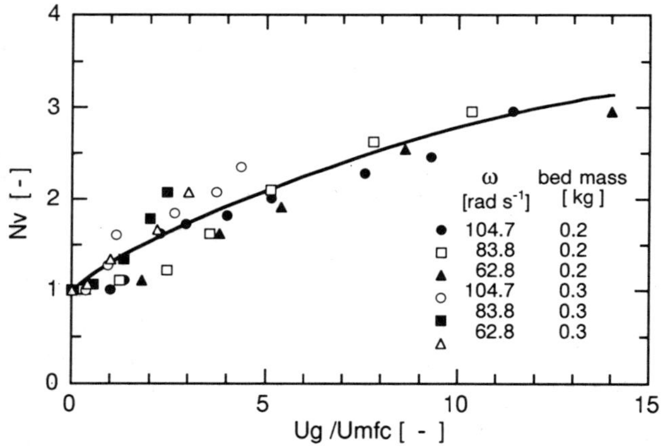

FIGURE 4. Bed expansion ratio versus dimensionless gas velocity.

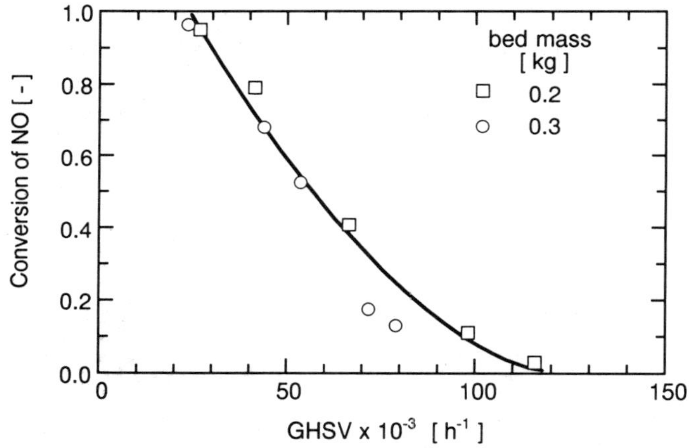

FIGURE 7. Effect of bed mass on NO conversion.

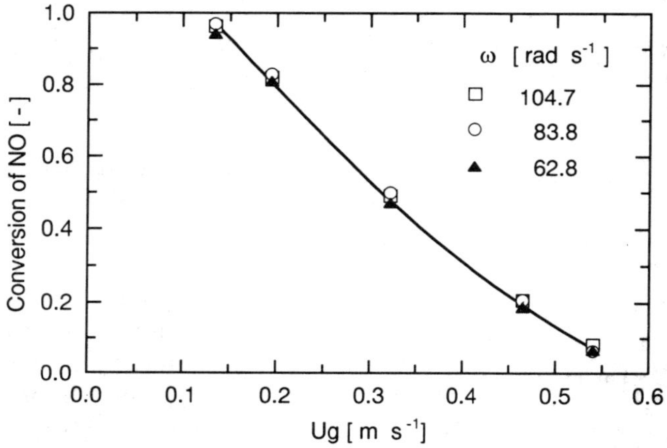

FIGURE 5. Effect of gas velocity on NO conversion.

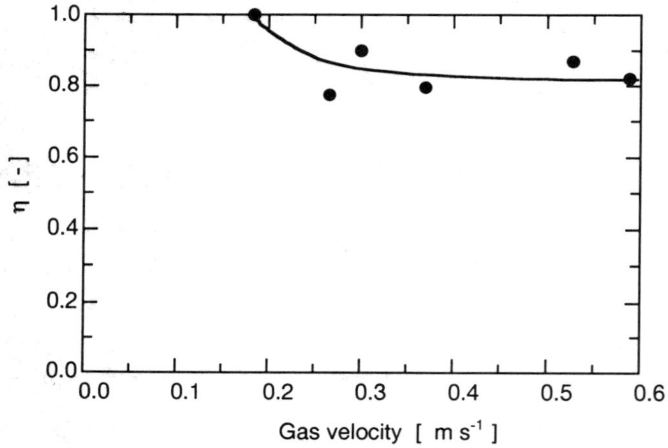

FIGURE 8. Effect of gas velocity on soot removal efficiency.

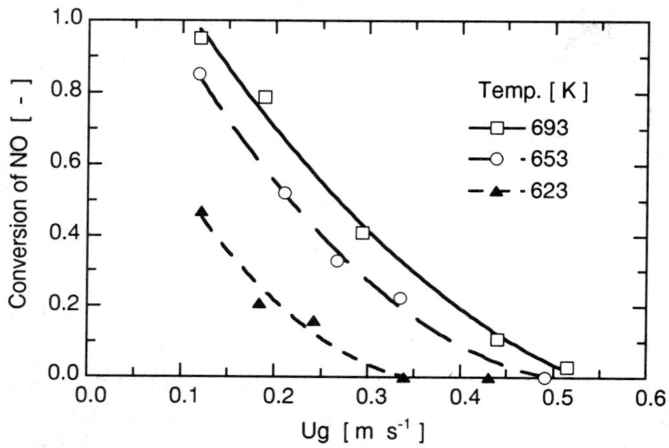

FIGURE 6. Effect of bed temperature on NO conversion.

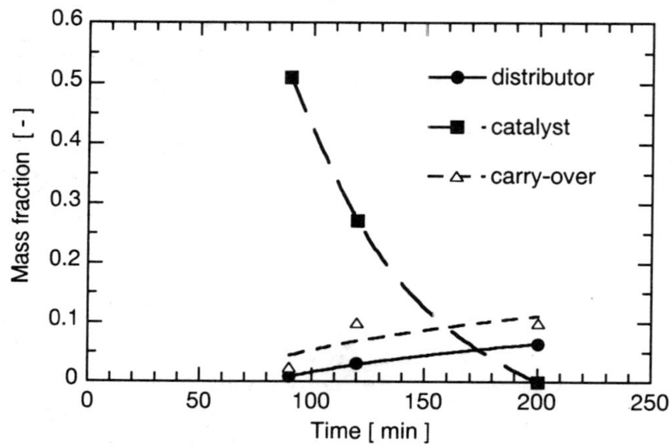

FIGURE 9. Mass fractions of soot versus time.

Fluidized and Moving Bed Desulfurization by Zinc Titanate

S. Dutta
Fluor Daniel Inc., Irvine, CA 92730

Both fluidized and moving bed reactor configurations are contenders for the hot gas clean-up technology~ using zinc titanate as the sorbent. Choosing the right configuration requires analysis due to many conflicting issues surrounding the process. Some of these issues are addressed in this work. For this study, only one type of fluidized bed, namely a circulating fluidized bed (CFB) is considered, in order to limit the scope of this study and to obtain a side-by-side comparison with a totally different kind of reactor, the moving bed (MB). Experimental rate data on cylindrical sorbent pellets, of the size range used in the MB, were available for a wide variety of operating conditions including high pressures. However, rate data on the same sorbent in the form of a fine powder were not available for use in the CFB projections. An attempt was therefore made to analyze the diffusional effects on the reaction system from the large particle data and project the fine powder behavior in the CFB. After determining the kinetic and diffusional characteristics of the selected sorbents, appropriate models for the two reactors are used to predict reactor performance. Effects of all major design and operating variables on the performance of the two reactors are compared.

Zinc titanate is a leading candidate for use as a regenerative sorbent for H_2S removal from hot gases, particularly in integrated gasification combined cycle(IGCC) power plants. Both fluidized and moving bed processes are contenders for hot gas clean-up using zinc titanate as the sorbent. Choosing the right reactor configuration, however, becomes difficult due to many conflicting issues surrounding the process. Composition, temperature and pressure of H_2S-bearing gases vary considerably depending on the source. Gases from oxygen and air-blown gasifiers typically contain 15-30% H_2 and 15-45% CO, compared to about 1% H_2 and 5% CO in gases from combustors operating under slightly substoichiometric conditions. The pressures and temperatures of the gases to be cleaned can also vary typically from 400 to 200 bar and 500 to 700 °C respectively depending on the source and heat recovery scheme.

For higher-pressure and lower-temperature applications a moving bed(MB) reactor is an obvious contender for this system, from mechanical considerations. Since sorbent losses due to reduction by H_2 and CO are known to be aggravated at higher temperatures, the lower-temperature MB reactor also appears attractive for gases containing high H_2 and CO. All these conditions are typical of an oxygen-blown gasifier.

On the other hand, for gases that contain low H_2 and CO, and are available at lower pressures and higher temperatures, a fluidized bed(FB) can be an attractive alternative. These gases are typical of combustors operating under slightly substoichiometric conditions.

The FB would normally use the sorbent in the form of a fine powder, instead of using large pellets as in the MB. Better predictability of sorbent behavior and sorbent utilization, due to minimum pore diffusional resistance, are the expected benefits of the FB system. There would be another key difference between the MB and FB systems if a circulating fluidized bed (CFB) is chosen. Due to a very high solid circulation rate that is characteristic of CFB's, the per-pass sorbent conversion would be negligible in the CFB. Due to this reason, only a very minor fraction of the total solid circulating through the absorber needs to be sent through the sorbent regenerator loop. In the MB absorber, the per-pass sorbent conversion would be substantial and 100% of the solid stream exiting the absorber would usually pass through the regenerator.

Several studies, for example, by Woods et al.[1] and Lew et al.[2, 4, 5], on the mechanism and kinetics of zinc titanate sorbents have appeared in recent years. However, no work has been published on the design aspects of this reaction system. The purpose of this work is to address some of the design aspects of the system. For this study, only one type of fluidized bed is considered among the various fluidized bed configurations to limit the scope of this study and to obtain a side-by-side comparison with a totally different kind of reactor like the MB. The type of fluidized bed chosen is a circulating fluidized bed(CFB).

A proper comparison between the MB and CFB systems, however, requires kinetic data on the same sorbent in two ranges of particle size. The first size range should correspond to that of the pellet sizes normally used in the MB, while the second range should correspond to the fine powders used in the CFB. Unfortunately no such data were available on the current reaction system. Therefore, an attempt has been made to analyze the diffusional effects on the reaction system to relate the large particle data obtained from one source with the fine powder data obtained form another. After establishing the kinetic and diffusional characteristics of the selected sorbents, appropriate models for the two reactors are used to predict the reactor behaviors.

REACTION KINETICS AND PORE DIFFUSIONAL EFFECTS

The following four reactions are considered as the principal reactions in a zinc titanate absorber :

$$ZnO\ (s) + H_2S\ (g) \Rightarrow ZnS\ (s) + H_2O\ (g) \quad (1)$$
$$CO\ (g) + H_2O\ (g) \Leftrightarrow H_2\ (g) + CO_2\ (g) \quad (2)$$
$$ZnO\ (s) + H_2\ (g) \Rightarrow Zn\ (s) + H_2O\ (g) \quad (3)$$
$$ZnO\ (s) + CO\ (g) \Rightarrow Zn\ (s) + CO_2\ (g) \quad (4)$$

The rate expressions used for the four reactions are as follows :

$$r_1 = k_{01} \exp(-E_1/RT)\ [ZnO][H_2S] \quad (5)$$
$$r_2 = k_{02} \exp(-E_2/RT)\ p_{H2}\ \{1 - p_{CO2}\, p_{H2}/(K_2 p_{CO} p_{H2O})\} \quad (6)$$
$$r_3 = k_{03} \exp(-E_3/RT)\ [ZnO][H_2] \quad (7)$$
$$r_4 = k_{04} \exp(-E_4/RT)\ [ZnO][CO] \quad (8)$$
$$K_2 = \exp(4578/T - 4.33)$$

The terms used in these and subsequent equations are explained in the list of notations. Rate constants for the desulfurization reaction (reaction 1) have been determined by a) simulating the TGA(Thermogravimetric Analysis) data on single cylindrical pellets (0.476 cm dia.) reported by Woods et al.[1], at the base case conditions shown in Table I, b) using a reaction activation energy of 42 kJ as suggested by Lew et al.[2], and c) accounting for pore diffusional effects according to the unreacted-core shrinking model of Wen[3].

Using the procedure proposed by Wen, the pore diffusional effects of the solid-gas reactions within pellets are determined by introducing an effectiveness factor, η_S , which is defined as the ratio of actual reaction rate to the rate without diffusional effects. Reaction rate without diffusional effects, or the intrinsic rate, is the rate determined by the chemical reaction rate expression of each solid-gas reaction as expressed by equation 5, 7 or 8, with gas concentrations in the bulk conditions. The effectiveness factor changes with solid conversion according to the following equation :

$$\eta_S = \frac{1}{1 + \phi_C \xi_C (1 - \xi_C) + \phi_C \xi_C^2 / N_{Sh}} \quad (9)$$

The Thiele modulus, ϕ_C , of Eqn.(9) is a function of effective diffusivity of the reacting gas, among other variables. Woods et al.[1] observed the reciprocal of effective diffusivity to be a linear function of pressure for the zinc titanate sorbents. Following that observation, the following correlation is used here to determine the effective diffusivity, D_{EA} :

$$1/D_{EA} = 1/D_{EA0} + 10(P-1) \quad (10)$$

where, $D_{EA0} = D\, \varepsilon_p/\tau$, and

$$1/D = 1/D_K + 1/D_M \quad (11)$$

where D_K and D_M are Knudsen and molecular diffusivities respectively. A value 10 is used for τ, the tortuosity factor, in the present simulation.

Rate Constants of ZnO Reductions

Rate constants for the ZnO reduction reaction by H_2 (reaction 3) are obtained by a) simulating the TGA data of Woods et al.[1] at the base case conditions shown in Table 1, and b) using an activation energy of 184 kJ as proposed by Lew et al.[4]. The same activation energy and half the rate constant compared to the H_2 reduction reaction are used for the CO reduction reaction (reaction 4), as also proposed by Lew et al.[4].

Rate Constants for Other Sorbents

The rate constants for the reduction reaction by H_2 for the four sorbent formulations considered in this study, namely $Z_{0.8}T$, ZT, Z_3T_2 and Z_2T, were determined by a) using the constants for Z_3T_2 as the basis and evaluating these constants by simulating the reduction rate data of Woods et al.[1] on the same sorbent formulation(Z_3T_2), and b) using the relative initial rates observed by Lew et al.[4] for zinc titanates of various ZnO contents. The initial reduction rates were found by Lew et al to increase with the increase in the ZnO content of the sorbent. The notations used here for the sorbent formulations indicate the ZnO to TiO_2 molar ratio in each sorbent. Thus $Z_{0.8}T$ stands for ZnO and TiO_2 in the molar ratio of 0.8 to 1.

Analysis of the experimental data of Woods et al. [1] indicates that the rate constants for the sulfidation reaction decrease with the increase in the ZnO content of the sorbent. Therefore, for this study, the sulfidation reation rate constants of $Z_{0.8}T$, ZT and Z_2T are determined from those of Z_3T_2 by assuming relative rates in an order opposite to that used for the reduction reaction by H_2.

It should be pointed here that the observations of Lew et al.[2, 5] indicate that the rate of sulfidation increases with the increase in ZnO content, a trend opposite to the observations reported by Woods et al.[1]. This discrepancy between the two observations could not be explained at the present time. Determination of rate parameters and the projections presented in this paper are based on the pellet data of Woods et al. because these data covered a very wide range of operating conditions of commercial interest, including the effects of pressure and feed H_2S concentration on the rates.

Intrinsic rate constants should ideally be determined from fine powder data which show negligible diffusional resistance. However, since no such data on fine powder covering a wide range of operating conditions were available, the pellet data of Woods et al. were analyzed with due care to extract the intrinsic rate constants and isolate the diffusional resistance at various operating conditions. The rate data of Woods et al. could not be simulated by the same intrinsic rate constant and varying physical properties and compositions of the sorbents. The intrinsic rate constants (the preexponential factors) were therefore also varied as shown in Table 1.

SIMULATION OF EXPERIMENTAL DATA

Rate constants, relevant particle properties and base case conditions used in the simulations are shown in Table 1. Simulation of the data on the effects of temperature, pressure, H_2S concentration and sorbent formulation, as observed in TGA experiments on single pellets(0.476 cm) by Woods et al.[1], is shown in Figures 1(a-d). In Figure 2(a), the TGA data of Lew et al.[5] showing the effect of temperature on fine powder(54 micron) are compared with the projected temperature effect on the same particle size with other conditions shown under base case in Table 1. In Figure 2(b) the experimental data and the projected curves for two pellet sizes (0.476 cm and 0.317 cm), with other conditions shown under base case in Table 1, are compared with the projection for 54 micron size and the data of Lew et al.[5] for the same size particle. These figures indicate that the simulation of experimental data on pellets can be considered satisfactory and use of the suggested rate parameters in MB projections would be appropriate. The projected rates for the fine powder (54 micron size) for the pellet material however are seen to be lower than those observed by Lew et al. for the material prepared and used in their laboratory. The CFB projections, based on the reaction rate of fine powders, presented in this paper, would therefore represent only the case of absorbent materials having same intrinsic reactivities as the pellet materials used in the MB projections.

REACTOR MODELS

Both CFB and MB reactor models are components of a general reactor model currently under development at Fluor Daniel that can apply to a large variety of solid-gas reaction and reactor systems. The CFB model is based on prediction of axial bed voidage profiles according to the correlations proposed by Kwauk et al.[6], and is described elsewhere (Dutta et al. [7]).

The non-catalytic solid-gas MB reactor model is an extension of the catalytic solid-gas fixed bed reactor model included in the general reactor model package. Contrary to once-through calculations required for other reactor models, however, simulation of the counter-current non-catalytic MB solid-gas reactor requires trial-and-error calculations. For this reactor, composition of the gas exiting from the top of the reactor and composition of the solid exiting from the bottom of the reactor are assumed. With these assumptions, step-wise calculations are carried out starting from the bottom of the reactor and continued until the top of the reactor is reached. The calculation is repeated until the calculated compositions of exit gas and solid streams match the assumed ones. Both CFB and MB models are one-dimensional, and effects of nonisothermality and back-mixing are ignored at the current stage.

KEY ISSUE OF CONCERN IN CFB ABSORBER DESIGN

Reaction rate in a CFB is a principal function of the local bed voidage. With other conditions remaining the same, the reaction rate becomes directly proportional to $1-\varepsilon$, where ε is the local bed voidage. Thus, for example, a three-fold increase in reaction rate would result with the decrease in bed voidage from 0.97 to 0.91. The key issue of concern in the CFB absorber design, therefore, is the ability to predict the bed voidage profiles as a function of the major design and operating variables. Five such variables are particle size, particle density, superficial gas velocity, solid circulation rate and bed pressure. Effects of four of these variables on axial bed voidage profiles, as predicted by the correlations by Kwauk et al.[6] for the current system, are shown in Figures 3(a-d). Improved correlations, and incorporation of radial profiles and core-annulus behaviors would lead to improved projections.

KEY ISSUE OF CONCERN IN MB ABSORBER DESIGN

The key issue of concern in the MB design is the ability to predict the effect of pore diffusion on the reaction rate for large particles. In this analysis, the effects are accounted for through the use of an effectiveness factor that changes with conversion. Effects of three major variables on the intraparticle effectiveness factor for the current system are shown in Figures 4(a-c). The figures clearly indicate that the actual reaction rate within a particle can drop to 5% or below (effectiveness factor 0.05) of the intrinsic reaction rate at higher temperatures and pressures for large particles. Figures 4(a-c) also point to a wide variation in the effectiveness factor, and hence the overall reaction rate of a particle, with the changes in design and operating conditions.

Figure 4(a) also shows the profile for a 54μ particle, typical of the size used in the CFB, that has an effectiveness factor close to unity. This illustrates the basic difference between the expected sorbent reactivity behaviors in the two absorber designs.

RESULTS

Projected effects of seven major design and operating variables on the MB absorber are compared side-by-side with those on the CFB absorber in Figures 5(a-g) and Figures 6(a-g). Each of these figures shows bed height as a function of per cent unconverted H_2S. The figures show the profiles for only the last 20% of the unconverted H_2S to highlight the differences in bed heights required for 99% H_2S removal targeted in this study. The points of 99% conversion are shown by symbols on the conversion profiles. Typical H_2S conversion profiles for the entire bed heights are shown in Figures 5(h) and 6(h) for both reactor configurations. All bed heights shown for the CFB include the height of a mixing zone 305 cm tall and 91 cm in diameter. Regenerated sorbent is mixed with recycle sorbent stream in this high-velocity mixing zone prior to the main reactor section of larger diameter. The sharp change in the conversion profiles shown in Figure 6(h) typically illustrates the effect of this mixing zone. The projections are for 20 tons per day of H_2S removal at the base case H_2S feed concentration of 0.5 mol%.

Base case conditions used in the projections are shown in Table 2. In the CFB, the feed is assumed to contain 50% ZnO and 17.07% ZnS, which results in a required slip-stream circulation rate of 1.9% to the CFB regenerator loop. The ZnO contents of sorbents Z_3T_2 and ZT, as shown in Table 2, are adjusted to contain the same weight fractions of ZnS and inert(8.38%) as in the Z_2T sorbent composition used in the base case. A gas composition typical of an oxygen-blown gasifier is chosen for the MB absorber, and that typical of a combustor operating under substoichiometric conditions is chosen for the CFB absorber. Typical gas composition from an air-blown gasifier would be intermediate between these two cases.

A common difference noticeable between the two sets of figures is a slower decay of the conversion profiles in the case of the CFB absorber. For the base case conditions the last 19% of H_2S capture takes more than half of the entire bed height. Low effective sorbent concentration per unit bed volume in CFB (due to high bed voidage) is responsible for this behavior. This indicates that for a much higher H_2S removal efficiency, say close to or above 99.9% , the counter current MB absorber would be preferred.

Effects of the major variables on the performance of these two reactor configurations are discussed below.

Effect of Temperature

The effect of temperature on H_2S capture is not very significant due to a low activation energy (42 kJ/mol) of the primary reaction and strong diffusional resistance in the case of the MB absorber. However, a higher temperature may lead to higher conversion to zinc and higher zinc loss by evaporation, as is discussed in a subsequent section.

Effect of Bed Diameter

As is obvious, a larger bed diameter, and hence a longer gas residence time, significantly improves the H_2S capture efficiencies of both MB and CFB absorbers. A larger bed diameter that results in a lower bed voidage further adds to the improved performance of the CFB. Mechanical constraints of pressure vessel design, channeling in MB, and hydrodynamic limitations of a CFB are among various factors that would dictate the maximum acceptable bed diameter.

Effect of Pressure

The effect of pressure is not seen to be very significant in the case of the MB absorber due to increased diffusional resistance at higher pressure as discussed earlier. The effect on the CFB absorber is also seen to be not very significant due to an apparent anomaly in the bed voidage projection at the highest pressure, as shown in Figure 3(d). According to this figure, the bed voidage near the top of the bed would be higher than that at the two lower pressures, in contrary to normal expectations. This indicates that pressure effect on CFB hydrodynamics needs to be more carefully evaluated for better projections.

Effect of Particle Size

The effect is very strong in the MB due to strong pore diffusional resistance indicated in this study. This study indicates that the rate of the primary reaction would be approximately inversely proportional to the particle diameter at the operating conditions. An opposite effect of particle size is indicated in the case of the CFB. The CFB absorber becomes increasingly more efficient with larger particle size due to lower bed voidage as predicted by the hydrodynamic model and shown in Figure 4(a).

Effect of Feed H_2S Concentration

Feed H_2S concentration is seen to have the strongest effect on the performance of a MB absorber, while it has no effect on the CFB. Due to the extremely high solid circulation rate, the sorbent concentration in a CFB remains essentially unchanged from inlet to outlet. The H_2S capture reaction thus becomes pseudo first order with respect to feed H_2S concentration and hence has no effect on conversion. This provides the CFB a distinct edge over the MB in being a versatile system capable of handling a wide range of H_2S feed concentration. It should be noted, however, that as the feed H_2S concentration changes, the slip-stream flowrate to the regenerator and the regenerator heat load need to be adjusted accordingly. The slip-stream flowrate required would be proportional to the feed H_2S concentration.

Effect of Sorbent Flowrate

Required sorbent circulation rate in the CFB absorber is about two orders of magnitude higher than that in the MB. The required bed height of the MB absorber is a strong function of the sorbent flowrate, that decreases with the increase of the sorbent flowrate. The effect is seen to be weak in the CFB absorber, again due to an apparent anomaly in the bed voidage projection at the highest sorbent flowrate, as shown in Figure 3(b). According to this figure, the bed voidage first decreases and then increases with the increase of the sorbent flowrate, with all other conditions remaining constant. This indicates that this aspect of the CFB hydrodynamics needs to be more carefully evaluated for better projections.

Effect of Sorbent Type

The rate constants for the three absorbents of this study increase in the order, Z_2T, Z_3T_2 and ZT, as shown in Table 1. The required bed heights for the MB are however in the reverse order, namely, highest for ZT and lowest for Z_2T. This is due to the opposing effect of the ZnO content (the reactive ingredient of the sorbent), which decreases in the order, Z_2T, Z_3T_2 and ZT.

The CFB performance is significantly affected by the bed voidage profile, which in turn is affected by the sorbent particle density. The particle densities decrease in the order, Z_2T, Z_3T_2 and ZT. According to the Figure 3(c), ZT, which has the lowest particle density, would have the highest bed voidage. Due to the opposing effects of rate constants and bed voidage the CFB performance is seen to be affected only slightly by the sorbent type.

Effects on Zinc Loss

The results shown in Figures 5 and 6 do not show the effects of the variables on the loss of sorbent due to the formation of zinc which could subsequently evaporate either in the absorber or in the regenerator at higher temperatures. The estimated zinc formation in the MB ranges from 0.15 to 3.00 tons per day in the ranges of operating conditions shown in Figures 5(a-g). The highest value of 3.00 tons per day was estimated for the sorbent Z_2T at the highest operating temperature of 873 K(600 °C) chosen in this study. The feed rate of zinc in the sorbent (in the form ZnO) is 66.7 tons per day at the base case conditions. Expected zinc losses from the moving bed process under the chosen operating conditions should therefore be considered substantial. The zinc formation is particularly sensitive to temperature at the high concentration levels of the reducing gases (H_2 and CO) used in the MB absorber. The zinc formation reduces to 0.15 tons per day with the decrease of temperature to 773 K(500 °C). The zinc formation rate reduces to 0.37 tons per day by switching to ZT at the base case temperature of 823 K(550 °C). These results indicate that sorbents with higher titanium contents and lower operating temperatures would be required to minimize the zinc loss in the MB process, for treating highly reducing gases.

Even with low levels of H_2 and CO (1% and 5% respectively) used in the CFB process, the estimated zinc formation is 1.62 tons per day using the sorbent Z_2T, at the highest operating temperature of 1023 K(750 °C) used for the CFB. Again with the decrease in the temperature to 923 K(650 °C), with other conditions remaining same, the zinc formation reduces to 0.288 tons per day. High titanium content absorbent and lower operating temperature therefore would also be required to minimize the zinc loss in the CFB system.

A fifth possible reaction, $Zn(g) + H_2S(g) \rightarrow ZnS(s)$, that may help in reducing the zinc loss is ignored in the present study in order to determine the maximum possible sorbent losses in both absorber and regenerator.

CONCLUSIONS

Proper design and operation of a MB zinc titanate absorber would require an accurate evaluation of the pore diffusional effects on the reaction rates. Proper design and operation of a CFB absorber, on the other hand, would require an accurate prediction of the bed voidage profiles as a function of the design and operating variables.

Apparent anomalies were observed on the hydrodynamic predictions of the CFB behavior. Experimental rate data on the same sorbent in the form of both pellets and fine powders was also not available. Despite this and assumptions made, the results presented should be satisfactory for the purpose of this work which is to make preliminary projections, compare the major differences in the expected performance of the two reactor configurations, and determine the right configuration for a given situation.

A counter-current MB reactor would be the preferred absorber configuration if a very high H_2S removal, say 99.9% or higher, is required. A CFB reactor, on the other hand, would be the preferred configuration for handling feed gas streams with widely varying H_2S contents. Mechanical constraints of pressure vessel design, channeling and hot spots in MB, and hydrodynamic limitations of a CFB are among various factors that would dictate the choice of the optimum absorber design.

Due to compactness and mechanical considerations, the MB would be the preferred configuration for use with oxygen-blown gasifiers. The CFB would be the choice, on the other hand, for use with combustors operating under slightly substoichiometric conditions.

Sorbent loss due to zinc formation, however, can be significant in both configurations, due to higher H_2 and CO contents in the feed gas in the first case(MB), and if higher operating temperatures are chosen in the second case(CFB). A proper combination of sorbent type and operating temperature would therefore be required to optimize the system performance in both cases.

ACKNOWLEDGMENT

Support of Fluor Daniel Inc. and permission to publish this work are gratefully appreciated. Introduction to this topic and encouragement of J. R. Joiner are particularly acknowledged.

NOTATIONS

a Mols of gas consumed per mol of solid reacted in each reaction
C_g, C_s Molar concentration of gas and solid, mol/cm^3
D_{EA0} Effective diffusivity, cm^2/s
D_M Molecular diffusivity, cm^2/s
D_K Knudsen diffusivity $(= \frac{19400\varepsilon p}{Sg\rho p}\sqrt{T/M})$, cm^2/s
d_p Particle diameter, cm
E_1-E_4 Activation energies of reactions 1-4, J/mol
$[H_2]$,[CO],
$[H_2S]$ Molar concentrations of the species, mol/cm^3
K_2 Equilibrium constant of reaction 2
k_{01}, k_{03},
k_{04} Preexponential factors of reactions 1, 3 & 4, based on external particle volume, cm^3/mol/s
k_{02} Preexponential factor of reaction 2, mol/bar/s
k_g Mass transfer coefficient, cm/s
k_s $= k_{ox} dp/6[\exp(-E_x/RT)]$
where k_{0x} is k_{01}, k_{03} or k_{04}, and E_x is E_1, E_3 or E_4, cm^4/mol/s
M Molecular weight of diffusing gas
m Order of reaction with respect to solid
n Order of reaction with respect to gas
N_{Sh} Sherwood number, $k_g r_p/D_{EA}$
P Pressure, absolute, bar
p_{H2}, p_{O2},
p_{CO}, p_{H2O}
Partial pressures of the species, bar
r_c Radius of unreacted shrinking core, cm
r_p Particle radius, cm
R Gas constant, J/mol/K
S_g BET pore surface area, m^2/g (cm^2/g in the expression for D_K)
S_{gi}, S_{gf} BET pore surface area, unreacted and totally converted, m^2/g
T Temperature, K
[ZnO] Molar concentration of ZnO in sorbent, mol/cm^3
ε Bed voidage of CFB
εp Sorbent porosity
ε_{pi}, ε_{pf}
Sorbent porosity, unreacted and totally converted
ηs Effectiveness factor based on surface
ξc r_c/r_p
ρp Particle density, g/cm^3
ϕc Thiele modulus $(= \frac{arpks(Cg)^{n-1}(C_s)^m}{D_{EA}})$
ϕp Particle shape factor

LITERATURE CITED

1. Woods, M.C., K.E. Leese, .K. Gangwal, D.P. Harrison and K. Jothimurugesan, "Reaction Kinetics and Simulation Models for Novel High-Temperature Desulfurization Sorbents" in DOE report #DOE/MC/24160-2571, 1989

2. Lew, S., A.F. Sarofim and M. F. Stephanopoulos, I&EC Res., 31(8), 1890, 1992

3. Wen, C. Y., I&EC, 60(9), 34, 1968

4. Lew, S., A.F. Sarofim and M.F. Stephanopoulos, Chem. Engg. Sci., 47(6), 1421, 1992

5. Lew, S., A.F. Sarofim and M.F. Stephanopoulos, AIChE J., 38(8), 1161, 1992

6. Kwauk, M.N., N.Wang, Y.Li and Z. Chen, "Fast Fluidization at ICM" in Circulating Fluid Bed Technology, ed. P. Basu, Pergamon Press, p.33, 1986

7. Dutta, S., B.R. Christian and M.F. Ratterman, "Circulating Fluid Bed Reactor Model Developed for FCC/ART Regenerator" in Fluidization VI, Engineering Foundation, p.9, 1989

Table 1
Rate Constants, Sorbent Particle Properties and Base Case Conditions Used in Simulations of Experimental Data

	0.47625 cm diameter Pellets				54 micron particles
	Base Case with Sorbent Z_3T_2	Sorbent $Z_{0.8}T$	Sorbent ZT	Sorbent Z_2T	
k_{01}	3.195E+05	7.668E+05	4.107E+05	2.396E+05	2.396E+05
k_{02}	1.253E+05	1.253E+05	1.253E+05	1.253E+05	1.253E÷05
k_{03}	2.953E+09	1.229E+09	2.294E+09	3.933E+09	3.933E+09
k_{04}	1.475E+09	6.145E+08	1.147E+09	1.967E+09	1.967E+09
E_1	41,868	41,868	41,868	41,868	41,868
E_2	122,003	122,003	122,003	122,003	122,003
E_3	184,219	184,219	184,219	184,219	184,219
E_4	184,219	184,219	184,219	184,219	184,219
d_p, cm	0.47625	0.47625	0.47625	0.47625	0.0054
$\rho p, g/cm^3$	1.730	1.628	1.663	1.807	1.807
ϕp	0.75	0.75	0.75	0.75	0.75
$\varepsilon_{pi}(\varepsilon_{pf})$	0.66 (0.48)	0.56 (0.41 @)	0.58 (0.42 @)	0.60 (0.44 @)	0.60 (0.44 @)
S_{gi} (S_{gf}), m^2/g	3.60 (5.60)	3.20 (5.20 @)	3.70 (5.70 @)	3.00 (4.70 @)	3.00 (4.70 @)
T, K	977	977	977	977	973
P, bar	1.013	1.013	1.013	1.013	1.013
Sorbent	Z_3T_2	$Z_{0.8}T$	ZT	Z_2T	Z_2T
Wt% ZnO	60.44	44.90	50.46	67.07	67.07
Feed Gas Compo. vol%					
H_2S	2.5	2.5	2.5	2.5	2.0
H_2O	15.0	15.0	15.0	15.0	0.0
CO	15.0	15.0	15.0	15.0	0.0
CO_2	5.0	5.0	5.0	5.0	0.0
H_2	10.0	10.0	10.0	10.0	1.0
N_2	Balance	Balance	Balance	Balance	Balance

@ Assumed values

Table 2
Base Case Conditions of MB and CFB Absorbers

	MB Absorber	CFB Absorber with Z_2T	CFB Absorber with Z_3T_2	CFB Absorber with ZT
Feed H_2S, tons/day	20.0	20.0	20.0	20.0
Feed gas flowrate, standard m^3/s	30.4	30.4	30.4	30.4
d_p, cm	0.635	0.0100	0.0100	0.0100
Bed Dia., cm	213.4	182.9	182.9	182.9
Mixing Zone Dia., cm	NA	91.4	91.4	91.4
Mixing Zone Ht., cm	NA	304.8	304.8	304.8
Absorbent Circulation Rate, tons/day	184	17600	17600	17600
T, K	823	973	973	973
P, bar	405	203	203	203
Sorbent	Z_2T	Z_2T	Z_3T_2	ZT
Wt% ZnO	67.07	50.00	45.06	37.62
Wt% ZnS	0.00	17.07	17.07	17.07
Feed Gas Composition , vol% :-				
H_2S	0.50	0.50	0.50	0.50
H_2O	12.38	25.00	25.00	25.00
CO	44.76	5.00	5.00	5.00
CO_2	9.82	25.00	25.00	25.00
H_2	30.38	1.00	1.00	1.00
N_2	Balance	Balance	Balance	Balance

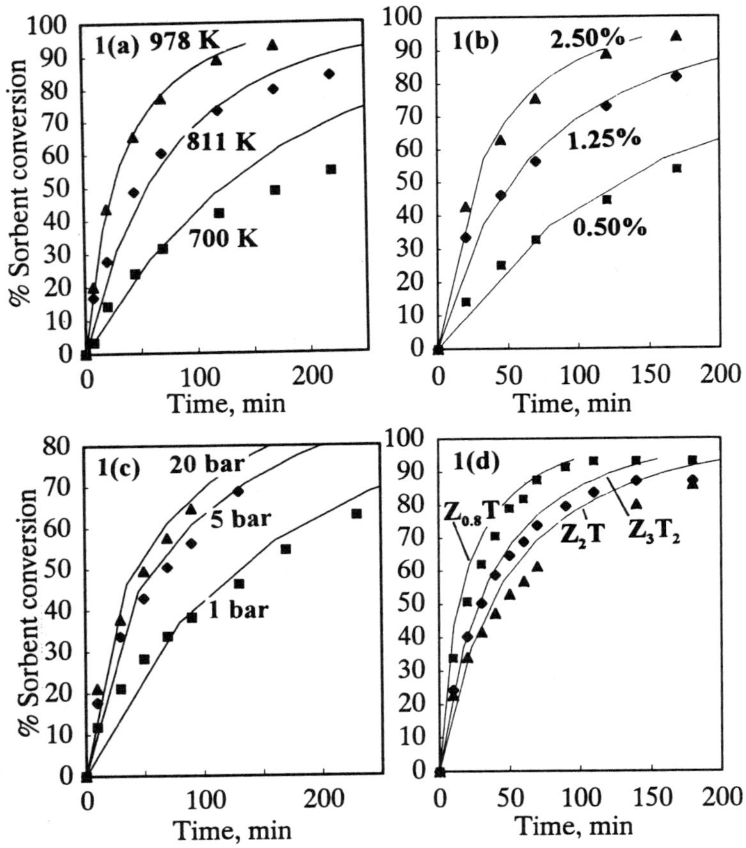

Figure 1. Simulation of TGA data of Woods et al. [1] on zinc titanate pellets, showing effects of a) temperature, b) H_2S concentration, c) pressure, and d) sorbent type

(Note : H_2S concentration in 2(c) is 0.50%)

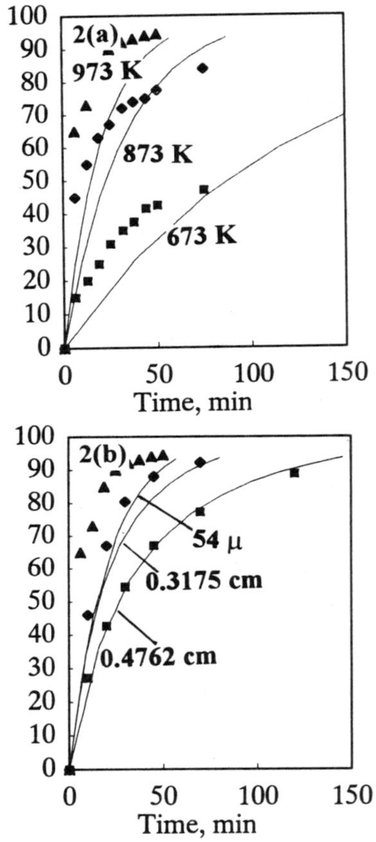

Figure 2 Comparison of fine powder data of Lew et al.[5] with the data and and projections for the base case sorbent, showing effects of a) temperature and b) particle size

(In Figure 2(b), only the data on 54μ size is from Lew et al.)

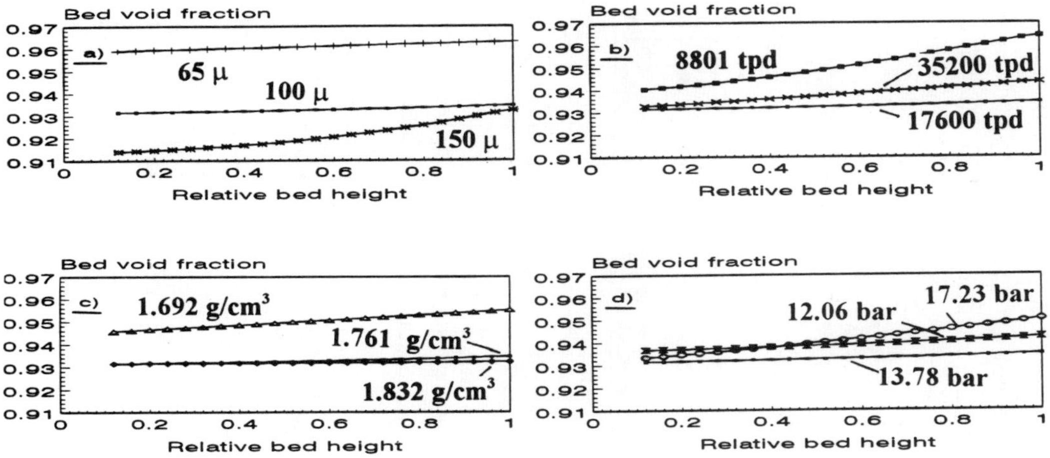

Figure 3. Estimated axial bed voidage profiles in the CFB absorber, showing effects of a) particle size, b) sorbent circulation rate, c) particle density and e) pressure

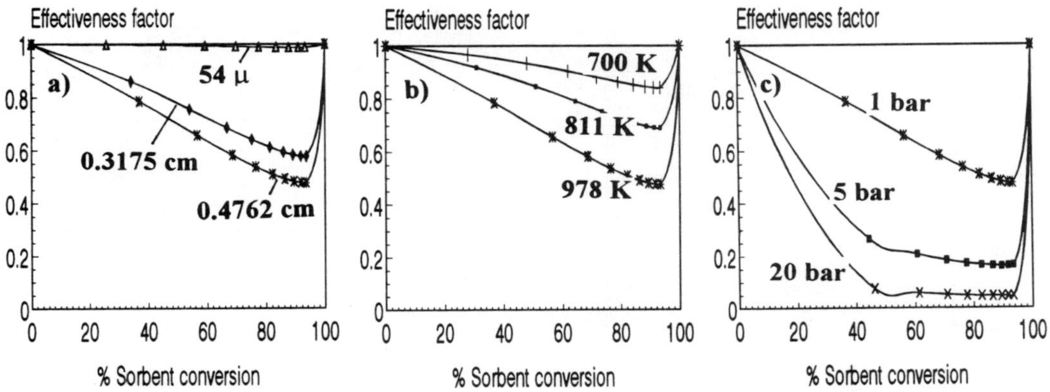

Figure 4. Estimated sorbent effectiveness factors as a function of a) pellet or particle size, b) temperature, and c) pressure

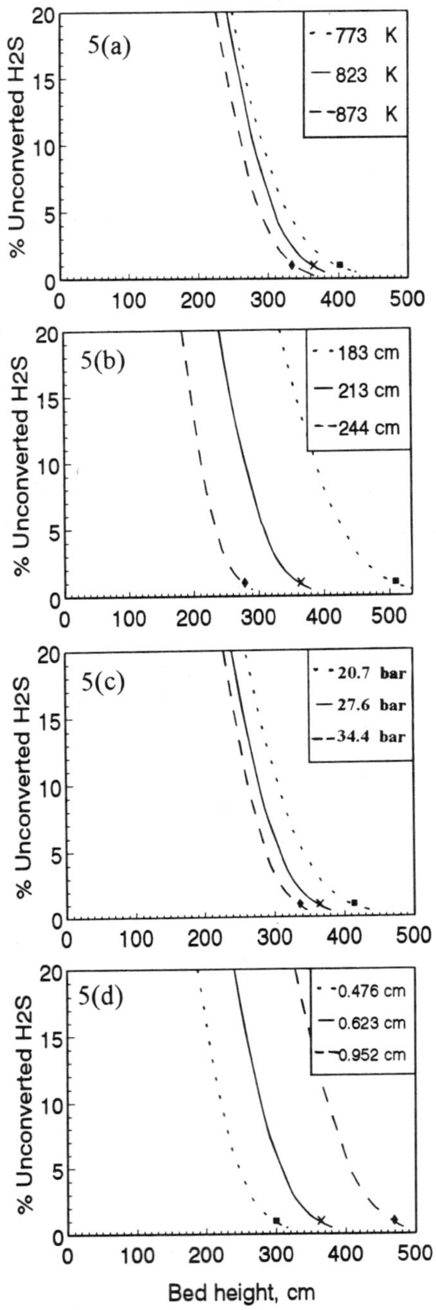

Figure 5. Projected effects of design and operating conditions on MB absorber

a) Effect of Temperature b) Effect of Bed Diameter
c) Effect of Pressure d) Effect of Particle Size

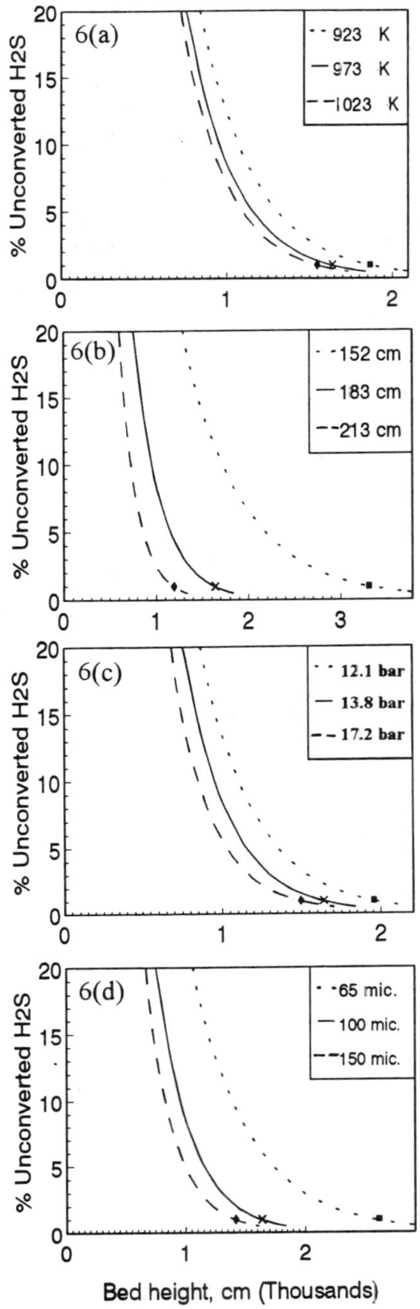

Figure 6. Projected effects of design and operating conditions on CFB absorber

a) Effect of Temperature b) Effect of Bed Diameter
c) Effect of Pressure d) Effect of Particle Size

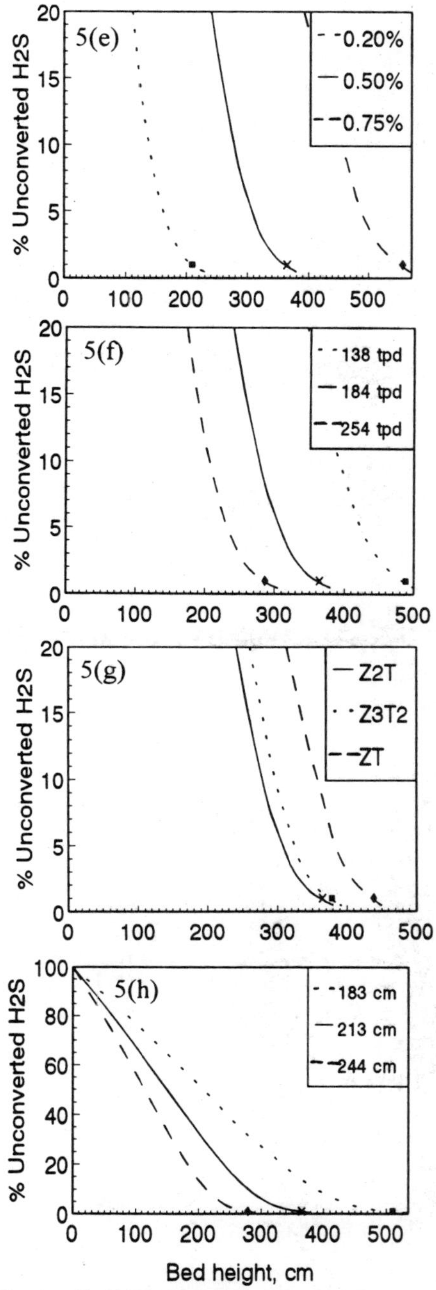

Figure 5. MB absorber projections contd.

e) Effect of H2S Concentration f) Effect of Sorbent Flowrate
g) Effect of Sorbent Type
h) Effect of Bed Diameter Showing profiles for entire bed height

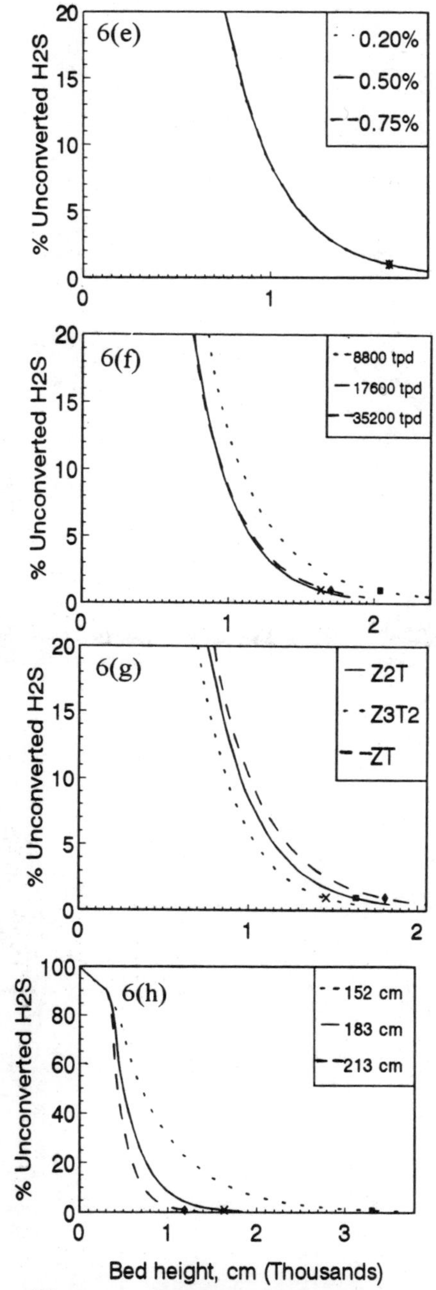

Figure 6. CFB absorber projections contd.

e) Effect of H2S Concentration f) Effect of Sorbent Flowrate
g) Effect of Sorbent Type
h) Effect of Bed Diameter Showing profiles for entire bed height

Index